From the Department of Cell Biology, University of Jyväskylä, Jyväskylä and the Department of Toxicology, National Public Health Institute, Kuopio, Finland

Kristian Jansson

Studies on the Genotoxicity of Chlorophenols in Mammalian Cells

Academic Dissertation

To be presented with the assent of the Faculty of Mathematics and Natural Sciences of the University of Jyväskylä for public examination in the Auditorium S212, on August 20, 1993, at 12 o'clock noon.

Kuopio 1993

Author's address:
Department of Toxicology
National Public Health Institute
P.O. Box 95
70701 Kuopio
Finland
Tel. +358 71 201211
Fax +358 71 201265

ISBN 952-90-4689-8

Kuopio University Printing Office
Kuopio 1993
Finland

ABSTRACT

Chlorophenols have been used widely as pesticides and as intermediates in industrial syntheses. They are formed indirectly as by-products in the chlorine bleaching of wood pulp, the chlorine disinfection of water, and the incineration of municipal waste. Furthermore, chlorophenols have been identified as mammalian metabolites of hexachlorocyclohexane isomers and chlorobenzenes.

In the present work, the genotoxicity of chlorophenols was studied with special reference to the mouse and rat carcinogen 2,4,6-trichlorophenol and the mouse carcinogen and somatic mutagen pentachlorophenol. The latter is known to be metabolized to tetrachlorohydroquinone in rodents and probably in humans. All experiments were performed in V79 Chinese hamster cells without exogenous metabolic activation.

Pentachlorophenol, 2,4- and 2,6-dichlorophenol, 2,4,5- and 2,4,6-trichlorophenol, and 2,3,4,6-tetrachlorophenol failed to induce mutation at the hypoxanthine phosphoribosyl transferase (*hprt*) locus to 6-thioguanine resistance. However, each of the 6 chlorophenols tested reduced the relative survival (reproductive integrity) in a dose-dependent manner.

Quantitative structure-activity relationships were subsequently derived for the toxicity of 9 chlorophenols (2-, 3-, and 4-chlorophenol, 2,4- and 2,6-dichlorophenol, 2,4,5- and 2,4,6-trichlorophenol, 2,3,4,6-tetrachlorophenol, and pentachlorophenol) in the survival assay. The toxicity was linearly related to both hydrophobicity and electronic (electron-withdrawing) substituent effects. These results suggest a relationship between the toxicity and the well known ability of chlorophenols to induce proton permeability in membranes.

2,4,6-Trichlorophenol induced dose-dependent increases in hyperdiploidy and micronuclei. However, it did not induce structural chromosome aberrations. These results suggest that 2,4,6-trichlorophenol causes chromosome malsegregation as its major mode of genotoxic action in mammalian cells. Further studies *in vivo* would be needed to elucidate the role of chromosome malsegregation in the carcinogenic action of 2,4,6-trichlorophenol.

Tetrachlorohydroquinone, the major rodent metabolite of pentachlorophenol, induced mutation at the *hprt* locus to 6-thioguanine resistance in a dose-dependent manner. This indicates that tetrachlorohydroquinone is a direct gene mutagen in mammalian cells and suggests that it is at least partly responsible for the *in vivo* mutagenicity and carcinogenicity of pentachlorophenol. Tetrachlorohydroquinone did not induce mutation at the *Na/K ATPase* locus to ouabain resistance. This locus specifity could be explained by the ability of tetrachlorohydroquinone to produce reactive oxygen species, which may cause large genetic damage such as deletions, resulting in mutation at the *hprt* but not at the *Na/K ATPase* locus. Tetrachlorocatechol, another metabolite of pentachlorophenol, did not induce mutation at either locus.

Tetrachlorohydroquinone induced dose-dependent increases in micronuclei. Moreover, the induction was partially inhibited by dimethyl sulfoxide, a hydroxyl radical scavenger. These results indicate that tetrachlorohydroquinone is a direct inducer of chromosomal damage in mammalian cells, and that hydroxyl radicals are at least partly responsible for this damage.

ACKNOWLEDGEMENTS

The work reported here was carried out mainly at the Department of Cell Biology, University of Jyväskylä, Jyväskylä, Finland, in collaboration with Vuokko Jansson, M.Sc.

I thank Dr. Marja Liisa Hattula for originally suggesting the study topic to me and Prof. Antti Arstila for placing the necessary facilities at my disposal.

The work was completed at the Department of Toxicology, National Public Health Institute, Kuopio, Finland. I am grateful to Dr. Jorma Mäki–Paakkanen for his support and advice.

Finally, I thank Drs. Kaija Linnainmaa and Hannu Norppa for reviewing the manuscript.

LIST OF ORIGINAL PUBLICATIONS

This thesis is based on the following original articles referred to in the text by their Roman numerals:

I Jansson K, Jansson V (1986) Inability of chlorophenols to induce 6–thioguanine–resistant mutants in V79 Chinese hamster cells. Mutation Res 171:165–168.

II Jansson K, Jansson V (1993) The toxicity of chlorophenols in V79 Chinese hamster cells. Toxicol Lett (in press).

III Jansson K, Jansson V (1992) Genotoxicity of 2,4,6–trichlorophenol in V79 Chinese hamster cells. Mutation Res 280:175–179.

IV Jansson K, Jansson V (1991) Induction of mutation in V79 Chinese hamster cells by tetrachlorohydroquinone, a metabolite of pentachlorophenol. Mutation Res 260:83–87.

V Jansson K, Jansson V (1992) Induction of micronuclei in V79 Chinese hamster cells by tetrachlorohydroquinone, a metabolite of pentachlorophenol. Mutation Res 279:205–208.

TABLE OF CONTENTS

1 BACKGROUND

1.1 General properties and sources of chlorophenols

Chlorophenols constitute a series of 19 congeners, ranging from monochlorophenols to the fully chlorinated pentachlorophenol. All of the chlorophenols are solids at room temperature except for 2–chlorophenol which is liquid. Chlorophenols, which are weak organic acids, can induce proton permeability in artificial lipid–bilayer membranes and uncouple ATP synthesis (oxidative phosphorylation) from the electron transport in isolated mitochondria (Stockdale and Selwyn, 1971; Miyoshi et al., 1987). Chlorophenols lack structurally alerting features which might identify them as potential electrophilic (DNA–reactive) mutagens or carcinogens (Ashby and Tennant, 1991).

Pentachlorophenol, and to a lesser extent, 2,3,4,6–tetrachlorophenol, and 2,4,6–trichlorophenol have been used as pesticides, primarily as fungicides for wood preservation; 2,4–dichlorophenol and 2,4,5–trichlorophenol have been used as intermediates in the manufacture of industrial and agricultural chemicals (IARC, 1986, 1991). Chlorophenols are formed indirectly as by–products in the chlorine bleaching of wood pulp, the chlorine disinfection of water, and the incineration of municipal waste (Paasivirta et al., 1985, 1992). In addition, chlorophenols have been identified as mammalian metabolites of hexachlorocyclohexane isomers (Angerer et al., 1983; Munir et al., 1984) and chlorobenzenes (Chu et al., 1984; Tanaka et al., 1986; Renner, 1988).

1.2 Biotransformation of chlorophenols

1.2.1 Chlorophenols other than pentachlorophenol

After intravenous administration of 2,4–dichlorophenol to rats, the metabolites found in urine were the glucuronide and other conjugates of the parent compound (Somani and Khalique, 1982). In rats dosed daily for 15 days with 2,4,6–trichlorophenol by gavage, the principal metabolite found in urine was the glucuronide conjugate of the parent compound (Bahig et al., 1981). In addition, two minor metabolites identified as the glucuronide conjugates of 2,3,6– and 2,4,5–trichlorophenol were isolated. After intraperitoneal administration of 2,4,6–trichlorophenol to rats, the only metabolites found in the blood were conjugates of the parent compound (Pekari et al., 1986). 2,3,5,6–Tetrachlorophenol was the only tetrachlorophenol metabolized in a significant degree after intraperitoneal administration to rats (Ahlborg and Larsson, 1978). The major metabolite found in urine after acid hydrolysis was tetrachlorohydroquinone.

1.2.2 Pentachlorophenol

In mice and rats given pentachlorophenol intraperitoneally, the metabolites found in urine were the glucurone conjugates of pentachlorophenol and tetrachlorohydroquinone, and free tetrachlorohydroquinone (Jacobson and Yllner, 1971, Ahlborg et al., 1974, 1978). The conversion of pentachlorophenol to tetrachlorohydroquinone was enhanced in rats by pretreatment with phenobarbital, 3–methylcholantrene, and 2,3,7,8–tetrachlorodibenzo–*p*–dioxin (Ahlborg and Thunberg, 1978). In rats given pentachlorophenol in feed for 30 or 107 days, the principal metabolite found in urine after acid hydrolysis was tetrachlorohydroquinone (Edgerton et al., 1979). In addition, 2,3,4,5–tetrachlorophenol, and tetrachlorocatechol were isolated. In a more recent study, pentachlorophenol and tetrachlorohydroquinone were found in urine as their sulfate and glucuronide conjugates as well as unconjugated molecules after oral administration of pentachlorophenol to rats (Reigner et al., 1991). In that study, total pentachlorophenol and tetrachlorohydroquinone accounted for 25 and 27% of the dose, respectively, when urine was collected for 72 h after administration.

Rat liver microsomes converted pentachlorophenol to tetrachlorohydroquinone (Ahlborg and Thunberg, 1978; Juhl et al., 1985; van Ommen et al., 1986), and also to tetrachlorocatechol (van Ommen et al., 1986). The conversion rate was enhanced by pretreatment of the rats with 2,3,7,8–tetrachlorodibenzo–*p*–dioxin, Aroclor 1254, hexachlorobenzene, phenobarbital, 3–methylcholantrene, and isosafrole (Ahlborg and Thunberg, 1978; Juhl et al., 1985; van Ommen et al., 1986).

Tetrachlorohydroquinone (Ahlborg et al., 1974; Edgerton et al., 1979) and also tetrachlorocatechol (Edgerton et al., 1979) have been detected in the urine of workers occupationally exposed to pentachlorophenol. In male volunteers, however, pentachlorophenol given orally was eliminated in urine as the parent compound and as the glucuronide conjugate (Uhl et al., 1986). No other metabolites were observed. It was suggested by Reigner et al. (1990) that the instability of tetrachlorohydroquinone in urine might explain inconsistent results concerning the occurrence of tetrachlorohydroquinone as a metabolite of pentachlorophenol in humans. In any case, human and rat liver microsomes showed similar activity in converting pentachlorophenol to tetrachlorohydroquinone (Juhl et al., 1985).

1.3 Genotoxicity of chlorophenols

1.3.1 Alteration of DNA base sequence

In *Salmonella typhimurium*, pentachlorophenol, the 6 dichlorophenols, 6 trichlorophenols, and 3 tetrachlorophenols did not induce mutation in tester strains TA98, TA100, TA1535, or TA1537 when tested in the absence of exogenous metabolic activation, or in the presence of liver S9 microsomes from Aroclor 1254–induced rats or Syrian hamsters (Haworth et al., 1983; Zeiger et al., 1988, 1992). However, 2,4,6–trichlorophenol and 2,3,4,6–tetrachlorophenol were found to induce mutation without exogenous metabolic activation at the X–linked hypoxanthine phosphoribosyl transferase (*hprt*) locus in V79 Chinese hamster cells (Hattula and Knuutinen, 1985).

In L5178Y mouse lymphoma cells, 2,4–dichlorophenol (Myhr et al., 1990) and 2,4,6–trichlorophenol (McGregor et al., 1988) induced mutation at the autosomal heterozygous thymidine kinase (*tk*) locus in the absence of exogenous metabolic activation. Pentachlorophenol induced recessive spots in the *in vivo* mouse somatic mutation test (Fahrig et al., 1978; reviewed by Seiler, 1991). Both of these mammalian systems are capable of detecting genetic alterations of several kinds, including chromosome aberrations, chromosome malsegregation, and mitotic recombination, in addition to gene mutation (Russel et al., 1981; Arlett, 1990).

1.3.2 Alteration of chromosome integrity or segregation

In Chinese hamster ovary (CHO) cells, 2,4–dichlorophenol (Anderson et al., 1990) and 2,4,6–trichlorophenol (Galloway et al., 1987) did not induce chromosome aberrations in the presence or absence of liver S9 from Aroclor 1254–induced rats, whereas pentachlorophenol (Galloway et al., 1987) induced chromosome aberrations only in the presence of the S9. Pentachlorophenol did not induce chromosome aberrations in cultured human peripheral lymphocytes when tested in the absence of exogenous metabolic activation (Ziemsen et al., 1987). The peripheral lymphocytes of pentachlorophenol–exposed workers showed a significant increase in chromosome–type aberrations (dicentrics and acentrics) compared with those of controls (Bauchinger et al., 1982). Pentachlorophenol did not induce sex–chromosome gain or loss in *Drosophila melanogaster* (Ramel and Magnusson, 1979).

1.3.3 Alteration of DNA integrity

In the absence of metabolic activation, pentachlorophenol did not induce DNA single-strand breaks in bacteriophage PM2 DNA, and it did not bind covalently to calf thymus DNA, either (Witte et al., 1985). However, pentachlorophenol bound covalently to calf thymus DNA in the presence of liver microsomes from hexachlorobenzene–induced rats (van Ommen et al., 1986). Tetrachlorohydroquinone induced DNA single–strand breaks in bacteriophage PM2 DNA and bound covalently to calf thymus DNA when tested in the absence of metabolic activation (Witte et al., 1985). Furthermore, tetrachlorohydroquinone induced DNA single-strand breaks in CHO cells in contrast to pentachlorophenol (Ehrlich, 1990). In the Microscreen prophage–induction assay in *Escherichia coli*, 7 of the 19 possible chlorophenol congeners, namely 2,3,4–, 2,3,6–, 2,4,5–, 2,4,6–, and 3,4,5–trichlorophenol, 2,3,4,5–tetrachlorophenol, and pentachlorophenol, induced prophage λ, but only in the presence of liver S9 from Arochlor 1254–induced rats (DeMarini et al., 1990).

1.3.4 DNA rearrangement

In CHO cells, 2,4,6–trichlorophenol (Galloway et al., 1987) did not induce sister–chromatid exchanges (SCEs) in the presence or absence of the S9, whereas pentachlorophenol (Galloway et al., 1987) and 2,4–dichlorophenol (Anderson et al., 1990) induced SCEs in the absence of exogenous metabolic activation. Pentachlorophenol did not induce SCEs in cultured human peripheral lymphocytes when tested in the absence of exogenous metabolic activation (Ziemsen et al., 1987), and there was no significant increase in SCEs in peripheral lymphocytes of smoking workers exposed to pentachlorophenol, as compared to smoking control subjects (Bauchinger et al., 1982). Peripheral lymphocytes from rats exposed to a commercial chlorophenolate formulation showed no increase in SCEs (Mustonen et al., 1989).

1.4 Carcinogenicity of chlorophenols

2,4,6–Trichlorophenol was found to be carcinogenic to rats and mice when administered in feed for 2 years (NCI, 1979). Dose–related increased incidences of lymphomas or leukemias were observed in male rats. The incidences of hepatocellular adenomas or carcinomas in male and female mice were significantly greater than those in controls. More recently, 2,4,6–trichlorophenol has been found to increase the rate of tumor progression in a transplant model for leukemia in rats (Dieter et al., 1989).

In a 2–year feed study of 2,4–dichlorophenol, there were no increased incidences of neoplastic lesions in dosed rats or mice of either sex compared with those in controls (NTP, 1989). Dose–related increased incidences of diffuse syncytial alteration of hepatocytes, characterized by individual hepatocytes with three or more nuclei, were observed in male mice.

Carcinogenesis studies of pentachlorophenol were conducted by feeding diets containing technical–grade pentachlorophenol or Dowiside EC–7 (a commercial grade with lower levels of contaminants) to male and female mice for 2 years (McConnell et al., 1991). Dose–related increased incidences of hepatocellular adenomas or carcinomas were observed in male and female mice exposed to either technical pentachlorophenol or Dowiside EC–7. The incidences of adrenal pheochromocytomas in male mice were significantly greater than those in controls. These neoplasms were also increased in female mice exposed to Dowicide EC–7. The incidences of hemangiosarcomas in the spleen or liver were significantly greater than those in controls for female mice that received technical–grade pentachlorophenol or Dowicide EC–7. It was concluded that the carcinogenic responses were due almost exclusively to penta–chlorophenol itself, with possibly a minimal potentiating influence by the contaminants in the induction of liver neoplasms in male mice.

2 AIMS OF THE STUDY

In the present work, the genotoxicity of chlorophenols was studied with special reference to the mouse and rat carcinogen 2,4,6–trichlorophenol and the mouse carcinogen and somatic mutagen pentachlorophenol. The latter is known to be metabolized to tetrachlorohydroquinone in rodents and probably in humans. All experiments were performed in V79 Chinese hamster cells without exogenous metabolic activation. The aims were as follows:

1) Examine the toxicity of chlorophenols and their ability to induce gene mutation

2) Find out the genotoxic profile of 2,4,6–trichlorophenol

3) Investigate the genotoxicity of tetrachlorohydroquinone, the major metabolite of pentachlorophenol

3 MATERIALS AND METHODS

3.1 Chemicals

Purified 2–, 3– and 4–chlorophenol, 2,4– and 2,6–dichlorophenol, 2,4,5– and 2,4,6–trichlorophenol, 2,3,4,6–tetrachlorophenol, pentachlorophenol and tetrachlorocatechol were kindly supplied by Dr. Juha Knuutinen (Department of Chemistry, University of Jyväskylä, Jyväskylä, Finland). The purity of these compounds was at least 99% by gas chromatography (Hattula and Knuutinen, 1985). Tetrachlorohydroquinone (purity at least 99% by titration) was purchased from Eastman Kodak (Rochester, NY, USA). Ethyl methanesulfonate was obtained from Sigma (St. Louis., MO, USA). Stock solutions of the chemicals were prepared immediately before use in acetone, dimethyl sulfoxide, or water, as described in original publications (I–V).

3.2 Cells

V79 Chinese hamster cells, originally obtained from Dr. Eliezer Huberman (then at the Oak Ridge National Laboratory, Oak Ridge, TN, USA), were cloned, checked for the presence of mycoplasma by the method of Chen (1977), and stored in ampoules frozen in liquid nitrogen. No HAT (hypoxanthine, aminopterin, thymidine) treatments were performed. This particular V79 subclone had a population doubling time of 14 h and a modal chromosome number of 22 (XY). Before each experiment, an ampoule of cells was thawed and maintained in exponential growth for 3 days. The cells were grown in Dulbecco's modified Eagle's medium (pH 7.4) obtained from Gibco (Paisley, UK) or K.C. Biological (Lenexa, KS, USA) and supplemented with 10% fetal calf serum, penicillin (100 IU/ml), and streptomycin (100 μg/ml) (all from Gibco) at 37°C with a humidified atmosphere of 10% CO_2 in air. Cell–culture dishes and flasks were purchased from Nunc (Roskilde, Denmark) or Falcon (Oxnard, CA, USA).

3.3 Chemical treatment

Cells were plated at 1×10^6 or 3×10^5 cells per 100–mm dish and incubated for 24 h. The cells were then treated for 24 h in growth medium (I–IV) or for 3 h in serum–free medium without antibiotics (V). For each dose level, 2–3 dishes were used. Each experiment included a solvent control and a positive control of ethyl methanesulfonate. All experiments were conducted without exogenous metabolic activation.

3.4 Assay procedures

3.4.1 *Toxicity (I–V)*

Toxicity was determined in a survival assay which gives a direct measure of reproductive cell death by measuring cloning efficiency. This method is routinely used in gene mutation assays and sometimes in conjugation with other tests on clonogenic cells (Scott et al., 1991).

After treatment, the cells were washed twice with Dulbecco's phosphate buffered saline (PBS), dissociated with 0.05% trypsin and 0.02% EDTA (Gibco) in PBS, and counted in a hemacytometer. To determine single–cell cloning efficiency, 200 cells were plated per 60–mm dish (4 dishes/dose), and after 6 days, the colonies were fixed in ethanol and stained with Giemsa. The survival was calculated as compared to the solvent control cloning efficiency.

3.4.2 *Gene mutation (I, III, IV)*

Forward mutations at the autosomal *Na/K ATPase* locus and the X–linked *hprt* locus were measured by resistance to ouabain and 6–thioguanine, respectively (Bradley et al., 1981; Nestmann et al., 1991). The essential *Na/K ATPase* locus permits the recovery of base–pair substitutions affecting only the ouabain–binding site of the Na/K ATPase enzyme, while the nonessential *hprt* locus permits the recovery of base–pair mutations, frameshift mutations, and small deletions (DeMarini et al., 1989).

After treatment, the cells were washed twice with PBS, dissociated, and counted, as described above. For phenotypic expression, 1–2 × 10^6 cells of each treatment condition were plated in a 100–mm dish and subcultured every 2 days, maintaining 1 × 10^6 cells at each subculture (I). Alternatively, 2 × 10^6 cells were plated in a 80–cm flask and subcultured every 2 days, maintaining 2 × 10^6 cells at each subculture (III, IV).

After a 6–day expression period, the cells were plated at 1 × 10^5 cells per 100–mm dish (10 dishes/dose), and 6–thioguanine (Sigma) was added 4 h later to give a final concentration of 60 μM (I). Alternatively, the cells were plated at 1 × 10^5 cells per 100–mm dish (20 dishes/dose) in medium containing 30 μM 6–thioguanine (III, IV) or at 2 × 10^5 cells per 100–mm dish (10 dishes/dose) in medium containing 1 mM ouabain (Sigma) (IV). The mutant colonies were fixed and stained 10 days later. To determine cloning efficiency at the time of mutant selection, 200 cells were plated per 60–mm dish (4 dishes/dose), and after 6 days, the colonies were fixed and stained. The mutant frequency was expressed as the number of mutants per 10^6 clonable cells.

3.4.3 *Micronuclei (III, V)*

Micronuclei arise from chromosome fragments or intact chromosomes excluded from the nucleus during cell division (Heddle et al., 1983). Therefore, the assay provides an indirect measure of the induction of chromosome aberrations or aneuploidy.

After treatment, the cells were washed twice with PBS, and dissociated as above after an additional 24–h incubation in fresh medium. The cells were suspended in a hypotonic solution of 75 mM KCl at 37°C for 10 min, fixed 3 times in ice–cold methanol:acetic acid (3:1), dropped onto wet glass slides, air dried, and stained with 5% Giemsa in 10 mM phosphate buffer (pH 6.8) for 20 min. The slides were mounted in DPX, coded, and 2000 interphase cells per dose (1000 cells/slide) were scored by one person for micronuclei using the criteria of Countryman and Heddle (1976).

3.4.4 Chromosome aberrations and hyperdiploidy (III)

After treatment, the cells were washed twice with PBS, and dissociated as above either immediately or after an additional 24-h incubation in fresh medium. Colcemid was added to a final concentration of 0.1 μg/ml 2 h before harvest. The cells were suspended in a hypotonic solution of 75 mM KCl at 37°C for 15 min, fixed 3 times in ice-cold methanol:acetic acid (3:1), dropped onto wet glass slides, air dried, and stained with 5% Giemsa in 10 mM phosphate buffer (pH 6.8) for 5 min. The slides were mounted in DPX and coded. For chromosome aberrations, 100 metaphase cells per dose (50 cells/slide) with 21-23 chromosomes were scored by one person. The aberrations excluding gaps and endoreduplications were grouped into categories of "simple" (breaks and terminal deletions), "complex" (exchanges and rearrangements), "other" (including pulverized chromosomes), and "total" as described by Galloway et al. (1987). For hyperdiploidy, 200 metaphase cells per dose (100 cells/slide) were scored by one person. Only diploid cells and hyperdiploid cells with 23-28 chromosomes were registered (Galloway and Ivett, 1986).

3.5 Data analysis

Quantitative structure-toxicity relationships (II) were analyzed by the Hansch method (Hansch and Fujita, 1969). An additive formula for the variance of mutant frequency (III) was adapted from Leong et al. (1985). Otherwise, the data were evaluated using standard parametric statistical methods as described in original publications (I-V).

4 RESULTS AND DISCUSSION

4.1 Gene mutation studies of chlorophenols (I, III)

Pentachlorophenol, 2,4– and 2,6–dichlorophenol, 2,4,5– and 2,4,6–trichlorophenol, and 2,3,4,6–tetrachlorophenol were tested without exogenous metabolic activation in V79 cells for the induction of forward mutation at the *hprt* locus to 6–thioguanine resistance. The chlorophenols reduced the relative survival (reproductive integrity) in a dose–dependent manner, but they failed to produce significant increases in the mutant frequency. The highest ineffective doses tested are shown in Table 1.

It has been previously reported by Hattula and Knuutinen (1985) that 2,4,6–trichlorophenol and 2,3,4,6–tetrachlorophenol induce mutation to 6–thioguanine resistance in V79 cells in the absence of exogenous metabolic activation. However, the numbers of cells they plated for expression and selection were insufficient for the responses they reported positive to be detected as statistically significant (Leong et al., 1985). The negative results presented here are in accordance with the negative *S. typhimurium* mutagenicity test results (Haworth et al., 1983; Zeiger et al., 1988) and the fact that chlorophenols have no structural features alerting to DNA reactivity (Ashby and Tennant, 1991).

Table 1. Highest ineffective doses (HID) of chlorophenols in the *hprt* gene mutation assay

Chlorophenol [1]	HID (μg/ml)	Relative survival at HID (%)	Ref.
2,4–Dichlorophenol	50	18	I
2,6–Dichlorophenol	500	17	I
2,4,5–Trichlorophenol	50	17	I
2,4,6–Trichlorophenol	100	53	I
2,4,6–Trichlorophenol	180	14	III
2,3,4,6–Tetrachlorophenol	100	10	I
Pentachlorophenol	50	27	I

[1] Treatment was for 24 h in growth medium.

4.2 Structure–toxicity relationships of chlorophenols (II)

The toxicity of 9 chlorophenols (2–, 3–, and 4–chlorophenol, 2,4– and 2,6–dichlorophenol, 2,4,5– and 2,4,6–trichlorophenol, 2,3,4,6–tetrachlorophenol, and pentachlorophenol) in terms of p*LC50*, *LC50* being the molar concentration required to reduce the relative survival to 50%, was analyzed quantitatively with physico–chemical parameters such as log *D*, *D* being the octanol–water distribution coefficient at pH 7.4, pK_a, K_a being the acid dissociation constant, and the Hammett substituent constant, σ. Eqns. 1–3 were derived by linear regression analysis:

$$\mathrm{p}LC50 = 0.818\ (\pm\ 0.492)\ \log D + 1.240\ (\pm\ 1.752) \qquad (1)$$
$$(n = 9,\ r^2 = 0.551,\ s = 0.374)$$

$$\mathrm{p}LC50 = 0.766\ (\pm\ 0.325)\ \log D - 0.215\ (\pm\ 0.105)\ \mathrm{p}K_a + 2.928\ (\pm\ 0.871) \qquad (2)$$
$$(n = 9,\ r^2 = 0.913,\ s = 0.178)$$

$$\mathrm{p}LC50 = 0.592\ (\pm\ 0.141)\ \log D + 0.890\ (\pm\ 0.165)\ \sigma + 1.262\ (\pm\ 0.358) \qquad (3)$$
$$(n = 9,\ r^2 = 0.985,\ s = 0.074)$$

In these expressions, n is the number of chlorophenols, r^2 is the correlation coefficient squared, and s is the standard deviation. The figures in parentheses are for construction of the 95% confidence limits.

Eqns. 1–3 show that the toxicity of chlorophenols is linearly related to both hydrophobicity, Log D, and electronic effects accounted for by $\mathrm{p}K_a$ or σ. The chlorophenols are known to be biologically active as uncouplers of mitochondrial oxidative phosphorylation because of their ability to induce proton permeability in membranes. This protonophoric activity of chlorophenols depends on both hydrophobicity and electron–withdrawing power (Stockdale and Selwyn, 1971; Miyoshi et al., 1987). Therefore, the toxicity of chlorophenols could result from their ability to induce proton permeability in mitochondrial or other cellular membranes.

Previous studies on cultured mammalian cells have suggested that chlorophenols disrupt cell function by nonspecific partitioning into hydrophobic cellular compartments rather than by being protonophores (Babich and Borenfreund, 1987; Aschmann et al., 1989; Seibert et al., 1989). However, those studies have measured chlorophenol–induced alterations in metabolic pathways or structural integrity which may not necessarily be related directly to cell death. By contrast, the survival assay used here gives a direct measure of reproductive cell death by measuring single–cell cloning efficiency.

4.3 Genotoxicity of 2,4,6–trichlorophenol (III)

Multiple genetic endpoints including mutation at the *hprt* locus to 6–thioguanine resistance, chromosome aberrations, hyperdiploidy, and micronuclei were measured in V79 cells after treatment with 2,4,6–trichlorophenol in the absence of exogenous metabolic activation.

2,4,6–Trichlorophenol was unable to induce 6–thioguanine–resistant mutants at doses up to 180 μg/ml, where the relative survival was reduced to 14% (Table 1). A qualitative reduction in the size of surviving colonies occurred at 90 μg/ml and higher. Thus, at these doses, 2,4,6–trichlorophenol affected cell proliferation as well as survival.

2,4,6–Trichlorophenol did not induce chromosome aberrations in cells fixed immediately after treatment. The highest dose scored was 60 μg/ml, where the mitotic index was reduced by about 78%. At this dose, 2,4,6–trichlorophenol did not induce chromosome aberrations in cells fixed 24 h after treatment, either. These results agree with those of Galloway et al. (1987), who showed that 2,4,6–trichlorophenol does not induce chromosome aberrations in CHO cells.

Table 2. Induction of hyperdiploidy and micronuclei by 2,4,6–trichlorophenol

Treatment [1] (μg/ml)	Hyperdiploid cells per 200 cells scored		Cells with micronuclei per 2000 cells scored	
	Expt. 1	Expt. 2	Expt. 1	Expt. 2
0	11	8	32	36
10	18	14	44	46
30	31 *	32 *	56 *	69 *
60	34 *	32 *	88 *	77 *
90	42 *	36 *	116 *	114 *

[1] Treatment was for 24 h in growth medium.
* Significantly different from control (χ^2 test, $P < 0.01$).

However, 2,4,6–trichlorophenol produced significant, dose–related increases in hyperdiploidy and micronuclei in cells fixed 24 h after treatment (Table 2). For both responses, the lowest effective dose was 30 μg/ml. At doses up to 60 μg/ml, there was no reduction in the mitotic index. It therefore appears that 2,4,6–trichlorophenol inhibits cell proliferation by a reversible mechanism rather than by causing cell killing.

The positive results for hyperdiploidy indicate that 2,4,6–trichlorophenol produces chromosome malsegregation leading to aneuploidy in V79 cells. Önfelt (1987) tested 4 chlorophenols (2–chlorophenol, 2,4– and 3,5–dichlorophenol, and 3,4,5–trichlorophenol) and found them all to induce spindle disturbances measured as c–mitosis in V79 cells in the absence of exogenous metabolic activation. In addition, Önfelt (1987) found the c–mitotic activity of these chlorophenols to be linearly related to their proton–donating (electron–withdrawing) ability. It has been suggested that the protonophoric activity of weakly acidic uncouplers, such as chlorophenols, might lead to spindle disturbances in mammalian cells (Önfelt, 1986). However, the precise mechanism is not known.

Micronuclei arise from chromosome fragments or intact chromosomes excluded from the nucleus during cell division (Heddle et al., 1983). 2,4,6–Trichlorophenol was unable to induce chromosome aberrations. Therefore, micronucleus formation by 2,4,6–trichlorophenol suggests malsegregational loss of whole chromosomes. 2,4,6–Trichlorophenol is a direct mutagen at the *tk* locus in L5178Y cells (McGregor et al., 1988). However, it was not mutagenic at the *hprt* locus in V79 cells (Table 1). Chromosome loss could be recovered at the heterozygous autosomal *tk* locus but not at the hemizygous X–linked *hprt* locus, which permits the recovery primarily of gene mutation (DeMarini et al., 1989; Liber et al., 1989; Arlett, 1990).

It should be noted that the genotoxic profile of 2,4–dichlorophenol, a definite non–carcinogen (NTP, 1989) is very similar to that of 2,4,6–trichlorophenol, a definite carcinogen (NCI, 1979). Like 2,4,6–trichlorophenol, 2,4–dichlorophenol is a direct mutagen at the *tk* locus in L5178Y cells (Myhr et al., 1990), is not a clastogen in CHO cells (Anderson et al., 1990), and was not mutagenic at the *hprt* locus in V79 cells (Table 1). In addition, 2,4–dichlorophenol has been shown to induce spindle disturbances measured as c–mitosis in V79 cells (Önfelt, 1987). Further studies *in vivo* of these two chlorophenols would be needed to elucidate the role of chromosome malsegregation in the carcinogenic action of 2,4,6–trichlorophenol.

4.4 Genotoxicity of pentachlorophenol metabolites

4.4.1 Gene mutation studies (IV)

Tetrachlorohydroquinone and tetrachlorocatechol, two metabolites of pentachlorophenol, were tested without exogenous metabolic activation in V79 cells for the induction of mutation at the *hprt* locus to 6–thioguanine resistance and at the *Na/K ATPase* locus to ouabain resistance. Tetrachlorohydroquinone produced significant, dose–related increases in the frequency of 6–thioguanine resistant mutants (Table 3). The lowest effective dose was 5 μg/ml, where the relative survival was about 63%. In contrast to tetrachlorohydroquinone, tetrachlorocatechol was unable to induce 6–thioguanine–resistant mutants at doses up to 40 μg/ml. The relative survival at this dose was about 5%. Both tetrachlorohydroquinone and tetrachlorocatechol were unable to induce ouabain–resistant mutants.

Pentachlorophenol has previously been shown to be a mouse carcinogen (McConnell et al., 1991) and a mouse somatic mutagen (Fahrig et al., 1978; reviewed by Seiler, 1991). However, tetrachlorohydroquinone, not pentachlorophenol (Table 1), was a direct mutagen at the *hprt* locus in V79 cells. This suggests that tetrachlorohydroquinone is at least partly responsible for the *in vivo* mutagenicity and carcinogenicity of pentachlorophenol.

In contrast to tetrachlorocatechol, tetrachlorohydroquinone readily autoxidizes at physiological pH values (Pritsos et al., 1987). As a result, tetrachlorohydroquinone is in redox equilibrium with the corresponding semiquinone radical and benzoquinone, the former of which is likely to be responsible for the ability of tetrachlorohydroquinone to bind covalently to DNA (Witte et al., 1985; van Ommen et al., 1986). On the other hand, the autoxidation of tetrachlorohydroquinone produces reactive oxygen species, which have been shown to be responsible for its ability to induce DNA strand breaks (Witte et al., 1985; Carstens et al., 1990).

In mammalian cells, the genotoxic effects of ionizing radiation and of radiomimetic chemicals such as bleomycin and streptonigrin are evidently mediated through reactive oxygen species, which cause DNA strand breaks leading to chromosome aberrations and deletion mutations (Hsie et al., 1986; Oya et al., 1986). Therefore, the mutagenic action of tetrachloro–hydroquinone could be explained by its ability to produce reactive oxygen species, which may cause large genetic damage such as deletions, resulting in mutation at the *hprt* locus to 6–thioguanine resistance but not at the *Na/K ATPase* locus to ouabain resistance. Note that the peripheral lymphocytes of pentachlorophenol–exposed workers showed induced chromosome aberrations qualitatively similar to those after exposure to ionizing radiation (Bauchinger et al., 1982).

Table 3. Induction of mutation to 6–thioguanine resistance at the *hprt* locus by tetrachlorohydroquinone

Treatment [1] (μg/ml)	Mutants per 10^6 clonable cells [2]		
	Expt. 1	Expt. 2	Average [3]
0	9 (13)	6 (8)	7
1	11 (13)	7 (9)	9
5	39 (45)	28 (36)	33 *
10	50 (66)	44 (61)	47 *
20	94 (112)	59 (67)	74 *

[1] Treatment was for 24 h in growth medium.
[2] Mutant frequency was calculated by dividing the number of mutant colonies (numbers in parentheses) by the number of cells plated for selection corrected by the cloning efficiency at the time of selection.
[3] Geometric mean.
* Significantly different from control (analysis of variance and Dunnett's test after log transformation, $P < 0.05$).

4.4.2 Micronucleus studies (V)

Tetrachlorohydroquinone was subsequently tested without exogenous metabolic activation in V79 cells for the induction of chromosomal damage measured as micronuclei. Table 4 shows that tetrachlorohydroquinone produced significant, dose–related increases in micronuclei in cells fixed 24 h after treatment. In both experiments, the lowest effective dose was 2.5 μg/ml, where the relative survival was about 62%.

Table 4. Induction of micronuclei by tetrachlorohydroquinone

Treatment [1] (μg/ml)	Cells with micronuclei per 2000 cells scored	
	Expt. 1	Expt. 2
0	29	34
1.25	53	45
2.5	89 *	110 *
3.75	142 *	151 *
5	195 *	163 *

[1] Treatment was for 3 h in serum–free medium.
* Significantly different from control (χ^2 test, $P < 0.01$).

Table 5. Inhibitory effect of dimethyl sulfoxide (DMSO) on the induction of micronuclei by tetrachlorohydroquinone (TCHQ) and ethyl methanesulfonate (EMS)

Treatment [1]	Cells with micronuclei per 2000 cells scored	
	Expt. 1	Expt. 2
DMSO (0.1%)	32	37
DMSO (5%)	26	34
EMS (1.5 mg/ml) + DMSO (0.1%)	124	131
EMS (1.5 mg/ml) + DMSO (5%)	109	138
TCHQ (3.75 μg/ml) + DMSO (0.1%)	133	145
TCHQ (3.75 μg/ml) + DMSO (5%)	69 *	84 *

[1] Treatment was for 3 h in serum–free medium.
* Significantly different from the corresponding treatment with 0.1% DMSO (χ^2 test, $P <$ 0.01).

Recently, Carstens et al. (1990) showed that the hydroxyl radical scavenger dimethyl sulfoxide at 5% effectively reduces the number of tetrachlorohydroquinone–induced DNA single–strand breaks in cultured human fibroblasts. It was therefore of interest to investigate if hydroxyl radicals were involved in the formation of micronuclei in V79 cells by tetrachlorohydroquinone.

As shown in Table 5, the induction of micronuclei by tetrachlorohydroquinone was significantly inhibited by dimethyl sulfoxide at 5%. This dimethyl sulfoxide concentration has previously been shown to protect V79 cells against the formation of DNA single–strand breaks by hydroxyl radicals from hydrogen peroxide (Ward et al., 1985). As expected, dimethyl sulfoxide had no effect on the induction of micronuclei by ethyl methanesulfonate, an alkylating agent (Table 5). This confirms that the inhibition of tetrachlorohydroquinone–induced micronucleus formation by dimethyl sulfoxide was due to its scavenging action. Therefore, hydroxyl radicals were at least partly responsible for the cytogenetic damage induced by tetrachlorohydroquinone.

5 SUMMARY AND CONCLUSIONS

In the present work, the genotoxicity of chlorophenols was studied with special reference to the mouse and rat carcinogen 2,4,6–trichlorophenol and the mouse carcinogen and somatic mutagen pentachlorophenol. The latter is known to be metabolized to tetrachlorohydroquinone in rodents and probably in humans. All experiments were performed in V79 cells without exogenous metabolic activation.

Pentachlorophenol, 2,4– and 2,6–dichlorophenol, 2,4,5– and 2,4,6–trichlorophenol, and 2,3,4,6–tetrachlorophenol reduced the relative survival (reproductive integrity) in a dose-dependent manner, but failed to induce mutation at the *hprt* locus to 6–thioguanine resistance. These results suggest that chlorophenols are not direct gene mutagens in mammalian cells. The chlorophenols tested in the present study have previously been shown to be unable to induce gene mutation in *S. typhimurium* (Haworth et al., 1983; Zeiger et al., 1988).

Quantitative structure–activity relationships were subsequently derived for the toxicity of 9 chlorophenols (2–, 3–, and 4–chlorophenol, 2,4– and 2,6–dichlorophenol, 2,4,5– and 2,4,6–trichlorophenol, 2,3,4,6–tetrachlorophenol, and pentachlorophenol) in the survival assay. The toxicity was linearly related not only to hydrophobicity, as suggested by previous mammalian–cell studies (Babich and Borenfreund, 1987; Aschmann et al., 1989; Seibert et al., 1989), but also to electronic (electron–withdrawing) substituent effects. These results suggest a relationship between the toxicity and the well known ability of chlorophenols to induce proton permeability in membranes.

2,4,6–Trichlorophenol induced dose–dependent increases in hyperdiploidy and micronuclei. However, it did not induce structural chromosome aberrations. These results suggest that 2,4,6–trichlorophenol causes chromosome malsegregation as its major mode of genotoxic action in mammalian cells. 2,4,6–Trichlorophenol has previously been shown by McGregor et al. (1988) to induce mutation at the heterozygous autosomal *tk* locus in L5178Y cells without exogenous metabolic activation. Further studies *in vivo* would be needed to elucidate the role of chromosome malsegregation in the carcinogenic action of 2,4,6–trichlorophenol.

Tetrachlorohydroquinone, the major rodent metabolite of pentachlorophenol, induced mutation at the *hprt* locus to 6–thioguanine resistance in a dose–dependent manner. This indicates that tetrachlorohydroquinone is a direct gene mutagen in mammalian cells and suggests that it is at least partly responsible for the *in vivo* mutagenicity and carcinogenicity of pentachlorophenol. Tetrachlorohydroquinone did not induce mutation at the *Na/K ATPase* locus to ouabain resistance. This locus specificity could be explained by the ability of tetrachlorohydroquinone to produce reactive oxygen species, which may cause large genetic damage such as deletions, resulting in mutation at the *hprt* but not at the *Na/K ATPase* locus. Tetrachlorocatechol, another metabolite of pentachlorophenol, did not induce mutation at either locus. This suggests that tetrachlorocatechol is not a direct gene mutagen in mammalian cells.

Tetrachlorohydroquinone induced dose–dependent increases in micronuclei. Moreover, the induction was partially inhibited by dimethyl sulfoxide, a hydroxyl radical scavenger. These results indicate that tetrachlorohydroquinone is a direct inducer of chromosomal damage in mammalian cells, and that hydroxyl radicals are at least partly responsible for this damage. It has previously been shown by Carstens (1990) that dimethyl sulfoxide reduces the number of tetrachlorohydroquinone–induced DNA single–strand breaks in human fibroblasts.

REFERENCES

Ahlborg UG, Larsson K (1978) Metabolism of tetrachlorophenols in the rat. Arch Toxicol 40:63–74.

Ahlborg UG, Larsson K, Thunberg T (1978) Metabolism of pentachlorophenol in vivo and in vitro. Arch Toxicol 40:45–53.

Ahlborg UG, Lindgren J–E, Mercier M (1974) Metabolism of pentachlorophenol. Arch Toxicol 32:271–281.

Ahlborg UG, Thunberg T (1978) Effects of 2,3,7,8–tetrachlorodibenzo–p–dioxin on the in vivo and in vitro dechlorination of pentachlorophenol. Arch Toxicol 40:55–61.

Anderson BE, Zeiger E, Shelby MD, Resnick MA, Gulati DK, Ivett JL, Loveday KS (1990) Chromosome aberration and sister chromatid exchange test results with 42 chemicals. Environ Mol Mutagen 16 (Suppl 18):55–137.

Angerer J, Maasz R, Heinrich R (1983) Occupational exposure to hexachlorocyclohexane: VI. Metabolism of γ–hexachlorocyclohexane in man. Int Arch Occup Environ Health 52:59–67.

Ashby J, Tennant RW (1991) Definitive relationships among chemical structure, carcinogenicity and mutagenicity for 301 chemicals tested by the U.S. NTP. Mutation Res 257:229–306.

Arlett CF (1990) Mammalian cell mutations. In Cooper CS, Grover PL (eds): "Chemical Carcinogenesis and Mutagenesis II." Springer–Verlag, Berlin, pp 27–50.

Aschmann C, Stork T, Wassermann O (1989) Short–term effects of chlorophenols on the function and viability of primary cultured rat hepatocytes. Arch Toxicol 63:121–126.

Babich H. Borenfreund E (1987) Structure–activity relationship (SAR) models established *in vitro* with the neutral red cytotoxicity assay. Toxicol In Vitro 1:3–9.

Bahig ME, Kraus A, Klein W (1981) Excretion and metabolism of 2,4,6–trichlorophenol–^{14}C in rats. Chemosphere 10:323–327.

Bauchinger MJ, Dresp J, Schmidt E, Hauf R (1982) Chromosome changes in lymphocytes after occupational exposure to pentachlorophenol (PCP). Mutation Res 102:83–88.

Bradley MO, Bhuyan B, Francis MC, Langenbach R, Peterson A, Huberman E (1981) Mutagenesis by chemical agents in V79 Chinese hamster cells: a review and analysis of the literature. A Report of the Gene–Tox Program. Mutation Res 87:81–142.

Carstens C–P, Blum JK, Witte I (1990) The role of hydroxyl radicals in tetrachloro–hydroquinone induced DNA strand break formation in PM2 DNA and human fibroblasts. Chem–Biol Interact 74:305–314.

Chen TR (1977) In situ detection of mycoplasma contamination by fluorescent Hoechst 33258 stain. Exp Cell Res 104:255–262.

Chu I, Villeneuve DC, Viau A, Barnes CR, Benoit FM, Qin YH (1984) Metabolism of 1,2,3,4–, 1,2,3,5–, and 1,2,4,5–tetrachlorobenzene in the rat. J Toxicol Environ Health 13:777–786.

Countryman PI, Heddle JA (1976) The production of micronuclei from chromosome aberrations in irradiated cultures of human lymphocytes. Mutation Res 41:321–332.

DeMarini DM, Brockman HE, de Serres FJ, Evans HH, Stankowski LF, Hsie AW (1989) Specific–locus mutations induced in eukaryotes (especially mammalian cells) by radiation and chemicals: a perspective. Mutation Res 220:11–29.

DeMarini DM, Brooks HG, Parkes DG (1990) Induction of prophage lambda by chlorophenols. Environ Mol Mutagenesis 15:1–9.

Dieter MP, Jameson CW, French JE, Gangjee S, Stefanski SA, Chhabra RS, Chan PC (1989) Development and validation of a cellular transplant model for leukemia in Fisher rats: a short term assay for potential anti–leukemic chemicals. Leukemia Res 13:841–849.

Edgerton TR, Moseman RF, Linder RE, Wright LH (1979) Multi–residue method for the determination of chlorinated phenol metabolites in urine. J Chromatogr 170:331–342.

Ehrlich W (1990) The effect of pentachlorophenol and its metabolite tetrachlorohydroquinone on cell growth and the induction of DNA damage in Chinese hamster ovary cells. Mutation Res 244:299–302.

Fahrig R, Nilsson C–A, Rappe C (1978) Genetic activity of chlorophenols and chlorophenol impurities. In Rao KR (ed): "Pentachlorophenol: Chemistry, Pharmacology, and Environmental Toxicology." Plenum Press, New York, pp 325–338.

Galloway SM, Armstrong MJ, Reuben C, Colman S, Brown B, Cannon C, Bloom AD, Nakamura F, Ahmed M, Duk S, Rimpo J, Margolin BH, Resnick MA, Anderson B, Zeiger E (1987) Chromosome aberrations and sister–chromatid exchanges in Chinese hamster ovary cells: evaluations of 108 chemicals. Environ Mol Mutagen 10 (Suppl 10):1–175.

Galloway SM, Ivett JL (1986) Chemically induced aneuploidy in mammalian cells in culture. Mutation Res 167:89–105.

Hansch C, Fujita T (1969) ρ–σ–π–Analysis. A method for the correlation of biological activity and chemical structure. J Am Chem Soc 86:1616–1626.

Hattula ML, Knuutinen J (1985) Mutagenesis of mammalian cells in culture by chlorophenols, chlorocatechols and chloroguaiacols. Chemosphere 14:1617–1625.

Haworth S, Lawlor T, Mortelmans K, Speck W, Zeiger E (1983) Salmonella mutagenicity test results for 250 chemicals. Environ Mutagen 5 (Suppl 1):3–142.

Heddle JA, Hite M, Kirkhart B, Mavournin K, MacGregor JT, Newell GW, Salamone MF (1983) The induction of micronuclei as a measure of genotoxicity: A Report of the U.S. Environmental Protection Agency Gene–Tox Program. Mutation Res 123:61–118.

Hsie AW, Recio L, Katz DS, Lee CQ, Wagner M, Schenley RL (1986) Evidence for reactive oxygen species inducing mutations in mammalian cells. Proc Natl Acad Sci USA 83:9616–9620.

International Agency for Research on Cancer (IARC) (1986) Occupational exposures to chlorophenols. In: "Some halogenated hydrocarbon and pesticide exposures." IARC Monographs on the Evaluation of the Carcinogenic Risk of Chemicals to Humans, Vol. 41,IARC, Lyon, pp 319–356.

International Agency for Research on Cancer (IARC) (1991) Pentachlorophenol. In: "Occupational exposures in insectiside application, and some pesticides." IARC Monographs on the Evaluation of Carcinogenic Risks to Humans, Vol. 53, IARC, Lyon, pp 371–402.

Jacobson I, Yllner S (1971) Metabolism of ^{14}C–pentachlorophenol in the mouse. Acta Pharmacol Toxicol 29:513–52

Juhl U, Witte I, Butte W (1985) Metabolism of pentachlorophenol to tetrachlorohydroquinone by human liver homogenate. Bull Environ Contam Toxicol 35:596–601.

Leong PM, Thilly WG, Morgenthaler S (1985) Variance estimation in single–cell mutation assays: comparison to experimental observations in human lymphoblasts at 4 gene loci. Mutation Res 150:403–410.

Liber HL, Yandell DW, Little JB (1989) A comparison of mutant induction at the *tk* and *hprt* loci in human lymphoblastoid cells: quantitative differences are due to an additional class of mutations at the autosomal *tk* locus. Mutation Res 216:9–17.

McConnel EE, Huff JE, Hejtmancik M, Peters AC, Persing R (1991) Toxicology and carcinogenesis studies of two grades of pentachlorophenol in B6C3F1 mice. Fundam Appl Toxicol 17:519–532.

McGregor HL, Brown A, Cattanach P, Edwards I, McBride D, Riach C, Caspary WJ (1988) Responses of the L5178Y tk^+/tk^- mouse lymphoma cell forward mutation assay: III. 72 coded chemicals. Environ Mol Mutagen 12:85–154.

Miyoshi H, Nishioka T, Fujita T (1987) Quantitative relationships between protonophoric and uncoupling activities of substituted phenols. Biochim Biophys Acta 891:194–204.

Munir KM, Nair J, Bhide SV (1984) Comparative formation of chlorophenol metabolites from hexachlorocyclohexane in mouse and rat in vivo and in vitro. Carcinogenesis 5:1519–1521.

Mustonen R, Elovaara E, Zitting A, Linnainmaa K, Vainio H (1989) Effects of commercial chlorophenolate, 2,3,7,8–TCDD, and pure phenoxyacetic acids on hepatic peroxisome proliferation, xenobiotic metabolism and sister chromatid exchange in the rat. Arch Toxicol 63:203–208.

Myhr B, McGregor D, Bowers L, Riach C, Brown AG, Edwards I, McBride D, Martin R, Caspary WJ (1990) L5178Y mouse lymphoma cell mutation assay results with 41 compounds. Environ Mol Mutagen 16 (Suppl 18):138–167.

National Cancer Institute (NCI) (1979) Bioassay of 2,4,6–trichlorophenol for possible carcinogenicity. NCI Technical Report No. 155. US Department of Health, Education, and Welfare, Public Health Service, National Institutes of Health, Bethesda, MD.

National Toxicology Program (NTP) (1989) Toxicology and carcinogenesis studies of 2,4–dichlorophenol in F344/N rats and $B6C3F_1$ mice. NTP Technical Report No. 353. US Department of Health and Human Services, Public Health Service, National Institutes of Health, Bethesda, MD.

Nestmann ER, Brillinger RL, Gilman JPW, Rudd CJ, Swierenga SHH (1991) Recommended protocols based on a survey of current practice in genotoxicity testing laboratories: II. Mutation in Chinese hamster ovary, V79 Chinese hamster lung and L5178Y mouse lymphoma cells. Mutation Res 246:255–284.

Önfelt A (1986) Mechanistic aspects on chemical induction of spindle disturbances and abnormal chromosome numbers. Mutation Res 168:249–300.

Önfelt A (1987) Spindle disturbances in mammalian cells: III. Toxicity, c–mitosis and aneuploidy with 22 different compounds. Mutation Res 182:135–154.

Oya Y, Yamamoto K, Tonomura A (1986) The biological activity of hydrogen peroxide: I. Induction of chromosome–type aberrations susceptible to inhibition by scavengers of hydroxyl radicals in human embryonic fibroblasts. Mutation Res 172:245–253.

Paasivirta J, Heinola K, Humppi T, Karjalainen A, Knuutinen J, Mäntykoski K, Paukku R, Piilola T, Surma–Aho K, Tarhanen J, Welling L, Vihonen H, Särkkä J (1985) Polychlorinated phenols, guaiacols and catechols in environment. Chemosphere 14:469–491.

Paasivirta J, Tenhola H, Palm H, Lammi R (1992) Free and bound chlorophenols in kraft pulp bleaching effluents. Chemosphere 24: 1253–1258.

Pekari K, Boudène C, Aitio A (1986) Kinetics of 2,4,6–trichlorophenol in different organs of the rat. Arch Toxicol 59:41–44.

Pritsos CA, Pointon M, Pardini RS (1987) Interaction of chlorinated phenolics and quinones with the mitochondrial respiration: a comparison of the *o*– and *p*–chlorinated quinones and hydroquinones. Bull Environ Contam Toxicol 38:847–855.

Ramel C, Magnusson J (1979) Chemical induction of nondisjunction in *Drosophila*. Environ Health Perspect 31:59–66.

Reigner BG, Gungon RA, Hoag MK, Tozer TN (1991) Pentachlorophenol toxicokinetics after intravenous and oral administration to rat. Xenobiotica 21:1547–1558.

Reigner BG, Rigod JF, Tozer TN (1990) Simultaneous assay of pentachlorophenol and its metabolite, tetrachlorohydroquinone, by gas chromatography without derivatization. J Cromatogr 533:111–124.

Renner G (1988) Hexachlorobenzene and its metabolism. Toxicol Environ Chem 18:51–78.

Russell LB, Selby PB, von Halle E, Sheridan W, Valcovic L (1981) Use of the mouse spot test in chemical mutagenesis: interpretation of past data and recommendations for future work. Mutation Res 86:355–379.

Scott D, Galloway SM, Marshall RR, Ishidate M, Brusick D, Ashby J, Myhr BC (1991) Genotoxicity under extreme culture conditions: A Report from ICPEMC Task Group 9. Mutation Res 257:147–204.

Seibert H, Kolossa M, Wassermann O (1989) Bovine spermatozoa as an *in vitro* model for studies on the cytotoxicity of chemicals: effects of chlorophenols. Cell Biol Toxicol 5:315– 330.

Seiler JP (1991) Pentachlorophenol. Mutation Res 257:27–47.

Somani SM, Khalique A (1982) Distribution and metabolism of 2,4–dichlorophenol in rats. J Toxicol Environ Health 9:889–897.

Stockdale M, Selwyn MJ (1971) Effects of ring substituents on the activity of phenols as inhibitors and uncouplers of mitochondrial respiration. Eur J Biochem 21:565–574.

Tanaka A, Sato M, Tsuchiya T, Adachi T, Niimura T, Yamaha T (1986) Excretion, distribution and metabolism of 1,2,4–trichlorobenzene in rats. Arch Toxicol 59:82–88.

Uhl S, Schmid P, Schlatter C (1986) Pharmacokinetics of pentachlorophenol in man. Arch Toxicol 58:182–186.

van Ommen B, Adang A, Müller F, van Bladeren PJ (1986) The microsomal metabolism of pentachlorophenol and its covalent binding to protein and DNA. Chem–Biol Interact 60:1–11.

Witte I, Juhl U, Butte W (1985) DNA damaging properties and cytotoxicity in human fibroblasts of tetrachlorohydroquinone, a pentachlorophenol metabolite. Mutation Res 145:71–75.

Zeiger E, Anderson B, Haworth S, Lawlor T, Mortelmans K (1988) Salmonella mutagenicity tests: IV. Results from the testing of 300 chemicals. Environ Mol Mutagen 11 (Suppl 12):1–158.

Zeiger E, Anderson B, Haworth S, Lawlor T, Mortelmans K (1992) Salmonella mutagenicity tests: V. Results from the testing of 311 chemicals. Environ Mol Mutagen 19 (Suppl 21):2–141.

Ziemsen B, Angerer J, Lehnert G (1987) Sister chromatid exchange and chromosomal breakage in pentachlorophenol (PCP) exposed workers. Int Arch Occup Environ Health 59:413–417.

I

Mutation Research, 171 (1986) 165–168
Elsevier

MTR 01084

Inability of chlorophenols to induce 6-thioguanine-resistant mutants in V79 Chinese hamster cells

Kristian Jansson * and Vuokko Jansson
Department of Cell Biology, University of Jyväskylä, Vapaudenkatu 4, SF-40100 Jyväskylä (Finland)

(Received 23 October 1985)
(Revision received 17 March 1986)
(Accepted 1 April 1986)

Summary

The induction of mutation at the hypoxanthine–guanine phosphoribosyl transferase locus and cytotoxicities of 6 different chlorophenols (2,4- and 2,6-dichlorophenol, 2,4,5- and 2,4,6-trichlorophenol, 2,3,4,6-tetrachlorophenol and pentachlorophenol) were examined in V79 Chinese hamster cells without exogenous metabolic activation. The chlorophenols were cytotoxic to V79 cells, but failed to produce significant increases in the frequency of 6-thioguanine-resistant mutants.

Chlorophenols are used extensively as wood preservatives (fungicides), bactericides, slimicides, herbicides and as intermediates in the manufacture of other pesticides. The annual world-wide production of chlorophenols has been estimated to be about 200 000 ton (Paasivirta, 1978). Chlorine bleaching of pulp produces chlorophenol-containing effluents (Lindström and Nordin, 1976). Furthermore, penta-chlorophenol and various di-, tri- and tetra-chlorophenols have been identified as mammalian metabolites of hexachlorocyclohexane (IARC, 1979; Munir et al., 1984).

In humans, soft tissue sarcoma, malignant lymphoma and leukemia have been associated with occupational exposure to 2,4,5-tri- and penta-chlorophenol, and to chlorophenol formulations containing mainly 2,4,6-tri-, 2,3,4,6-tetra- and penta-chlorophenol (IARC, 1982). In animal studies, 2,4,6-trichlorophenol (96–97% pure) has been shown to produce lymphomas and leukemias in male Fisher rats, and hepatocellular carcinomas or adenomas in both male and female B6C3F1 mice (National Cancer Institute, 1979). Long-term bioassays of pentachlorophenol by the U.S. National Toxicology Program are in progress (Huff, 1982).

The genotoxicity of chlorophenols has not been well characterized. 2,4,6-Trichlorophenol was reported to exhibit high DNA-damaging potency without exogenous metabolic activation in *Bacillus subtilis* rec-assay (Kinae et al., 1981). 2,4-Di- and 2,4,5-tri-chlorophenol induced stickiness and lagging of chromosomes, and chromosome fragmentation in *Vicia faba* (Amer and Ali, 1974). In *Saccharomyces cerevisiae*, in the absence of exogenous activation, 2,6-di- and 2,4,5-tri-chlorophenol failed to induce reverse mutation or mitotic gene conversion (Nestmann and Lee, 1983), whereas 2,4,6-trichlorophenol induced forward mutation but did not induce mitotic gene conversion, and pentachlorophenol induced both forward mutation and mitotic gene conversion (Fahrig et al., 1978). Pentachlorophenol failed to induce sex-linked recessive lethals in *Drosophila melanogaster* (Vogel and Chandler, 1974), and there was con-

* To whom all correspondence should be addressed.

flicting evidence with regard to the production of chromosomal aberrations in workers exposed to pentachlorophenol (Wyllie et al., 1975; Bauchinger et al., 1982).

The mutagenicity of chlorophenols in mammalian cells has not been determined. However, the chlorophenol derivative 2,4-D (2,4-dichlorophenoxyacetic acid) was shown to be mutagenic without exogenous activation in V79 Chinese hamster cells (Ahmed et al., 1977). To extend the previous genotoxicity studies, we examined the induction of mutation to 6-thioguanine resistance in V79 Chinese hamster cells by 6 different chlorophenols.

Materials and methods

Chemicals

2,4- and 2,6-dichlorophenol, 2,4,5- and 2,4,6-trichlorophenol, 2,3,4,6-tetrachlorophenol and pentachlorophenol were purchased from Fluka AG, Buchs, Switzerland, purified by recrystallization and determined to be > 99.5% pure by GC. Chlorinated dibenzo-*p*-dioxins were not detected (limit of detection, 0.05 ppm). Ethyl methanesulfonate (EMS) was obtained from Sigma Chemical Co., St. Louis, MO. Stock solutions of the chlorophenols (50 mg/ml in acetone) and EMS (50 mg/ml in water) were prepared immediately before use.

Cell culture

V79 Chinese hamster cells, originally obtained from Dr. Eliezer Huberman (Biology Division, Oak Ridge National Laboratory, Oak Ridge, TN), were cloned to reduce the spontaneous background frequency of 6-thioguanine-resistant mutants and stored in ampoules frozen in liquid nitrogen. No HATG (hypoxanthine, aminopterin, thymidine, glycine) treatments were performed. Prior to each experiment, an ampoule of cells was thawed and used. The cells were grown in Dulbecco's modified Eagle's medium (K.C. Biological, Inc., Lenexa, KS) supplemented with 10% fetal calf serum (Gibco Ltd., Paisley, Scotland), 2 mM glutamine (K.C. Biological), 100 IU/ml penicillin, 100 μg/ml streptomycin and 250 ng/ml fungizone (Gibco Ltd.) at 37°C with a humidified atmosphere of 10% CO_2 in air. Cell culture dishes were purchased from Falcon Co., Oxnard, CA.

Mutation assay

The forward mutation assay on 6-thioguanine resistance in V79 cells was based on the replating method described by McMillan and Fox (1979). 1×10^6 cells were plated in 100-mm dishes in 9 ml of medium and cultured for 24 h. The test chemical was then added to the dishes in 1 ml of medium to give the desired final concentrations. For each concentration, two dishes were used. Each experiment included a solvent control and a positive control of EMS at 200 μg/ml. The maximum solvent concentration was 1%. This level had no effect on cell growth or spontaneous mutant frequency. After 24 h of treatment, cells were dissociated with a trypsin–EDTA solution, and cell survival was determined by plating 200 cells in 60-mm dishes (4 dishes/dose) in 4 ml of medium and staining the dishes with Giemsa 6 days later. Colonies of at least 50 cells were counted and survival was calculated relative to the solvent control.

At the time cells were plated for survival, 1–2 $\times 10^6$ cells of each treatment condition were plated in 100-mm dishes in 10 ml of medium to allow phenotypic expression. Cells were subcultured every 48 h, maintaining 1×10^6 cells at each subculture. After a 6-day expression time, which is considered to be adequate in general (Bradley et al., 1981), 1×10^5 cells were plated in 100-mm dishes (10 dishes/dose) in 9 ml of medium, and 6-thioguanine (Sigma Chemical Co.) was added in 1 ml of medium 4 h later to give a final concentration of 10 μg/ml. The number of mutant colonies was determined 10 days later, after Giemsa staining. To determine cell viability at the time of mutant selection, 200 cells were plated in 60-mm dishes (4 dishes/dose) in 4 ml of medium, and the colonies were stained with Giemsa and counted 6 days later. Mutant frequencies were expressed as the numbers of mutants per 10^6 viable cells.

A chemical was considered to be mutagenic if the mutant frequency exceeded the 99% upper confidence limit of the historical background mutant frequency and a dose-related increase in the mutagenicity was observed. The historical background mutant frequency (the solvent control frequencies in 12 independent experiments including those reported here) was $7.0 \times 10^{-6} \pm 4.5 \times 10^{-6}$ (mean ± standard deviation). The 99% confidence interval calculated by multiplying the standard deviation by the appropriate t value of the Student's t distribution (one-sided) requiring 99% confidence was 12.2×10^{-6}, and the 99% upper confidence limit was 19.2×10^{-6}.

Results and discussion

The results are presented in Table 1. Each of the 6 chlorophenols tested reduced the plating efficiency (% cell survival) in a dose-dependent manner. This cytotoxic effect can be attributed to the ability of pentachlorophenol and other chlorophenols to inhibit oxidative phosphorylation (Farquharson et al., 1958; Weinbach et al., 1965). At the concentrations tested, none of the chlorophenols produced significant increases in the frequency of 6-thioguanine-resistant mutants. Evidence that the cells in these experiments were responsive was shown by EMS at 200 μg/ml. In a separate experiment, EMS induced marked dose-related increases in 6-thioguanine resistance.

There have been few studies regarding the induction of gene mutation by chlorophenols. Our negative results are in accordance with the findings of Nestmann and Lee (1983), who showed that 2,6-di- and 2,4,5-tri-chlorophenol were inactive in inducing reverse mutation in *S. cerevisiae*, and of Vogel and Chandler (1974), who reported that pentachlorophenol failed to induce sex-linked recessive lethals in *D. melanogaster*. Fahrig et al. (1978) found that 2,4,6-tri- and penta-chlorophenol induced forward mutation to cycloheximide resistance in *S. cerevisiae* without exogenous metabolic activation. However, the biochemical basis of cycloheximide resistance is poorly understood, and this mutation system is not considered to be completely reliable (Zimmermann et al., 1984).

Most human and animal carcinogens have been found to cause gene mutation, chromosome breakage or other types of DNA lesions. We point out that the apparent lack of induction of gene mutation by chlorophenols does not necessarily mean the lack of genotoxicity of these chemicals. In this respect, it is interesting to note the findings of

TABLE 1

INDUCTION OF MUTATION TO 6-THIOGUANINE (TG) RESISTANCE IN V79 CHINESE HAMSTER CELLS BY TREATMENT WITH CHLOROPHENOLS FOR 24 h

Chemical	Dose (μg/ml)	Survival (%) [a]	Number of TG-resistant mutants/10^6 viable cells
2,4-Dichlorophenol	0	100	12
	12.5	108	6
	25	80	6
	50	18	0
EMS [b]	200	84	965
2,6-Dichlorophenol	0	100	4
	125	88	0
	250	82	0
	500	17	6
EMS	200	62	584
2,4,5-Trichlorophenol	0	100	9
	6.25	90	3
	12.5	63	6
	25	45	5
	50	17	7
EMS	200	64	1262
2,4,6-Trichlorophenol	0	100	2
	12.5	85	3
	25	79	4
	50	72	2
	100	53	0
EMS	200	79	631
2,3,4,6-Tetrachlorophenol	0	100	1
	12.5	101	7
	25	88	16
	50	34	7
	100	10	9
EMS	200	88	1181
Pentachlorophenol	0	100	16
	6.25	90	10
	12.5	73	0
	25	53	0
	50	27	7
EMS	200	55	966
EMS	0	100	8
	50	74	180
	100	73	259
	200	53	613
	400	13	1367

[a] The average absolute plating efficiency of control cells was 75%, range 62–89%.

[b] Positive control, ethyl methanesulfonate.

Bauchinger et al. (1982), who reported that peripheral lymphocytes of pentachlorophenol-exposed workers showed a significant increase in chromosome-type aberrations (dicentrics and acentrics) but not in chromatid-type aberrations, such as breaks or exchanges. Further studies will be required to clarify the genotoxic properties of chlorophenols, as well as their suspected human and animal carcinogenicity.

Acknowledgements

We wish to thank Dr. Juha Knuutinen for purifying the chlorophenols and performing the GC analyses. We also wish to thank Mrs. Irene Helkala for technical assistance. This work was supported by grants from the Foundation of Maj and Tor Nessling, and the National Academy of Finland.

References

Ahmed, F.E., N.J. Lewis and R.W. Hart (1977) Pesticide induced ouabain resistant mutants in Chinese hamster V79 cells, Chem.-Biol. Interact., 19, 369–374.

Amer, S.M., and E.M. Ali (1974) Cytological effects of pesticides, V. Effects of some herbicides on *Vicia faba*, Cytologia (Tokyo), 39, 633–643.

Bauchinger, M., J. Dresp, E. Schmid and R. Hauf (1982) Chromosome changes in lymphocytes after occupational exposure to pentachlorophenol (PCP), Mutation Res., 102, 83–88.

Bradley, M.O., B. Bhuyan, M.C. Francis, R. Langenbach, A. Peterson and E. Huberman (1981) Mutagenesis by chemical agents in V79 Chinese hamster cells: A review and analysis of the literature, A report of the Gene-Tox Program, Mutation Res., 87, 81–142.

Fahrig, R., C.-A. Nilsson and C. Rappe (1978) Genetic activity of chlorophenols and chlorophenol impurities, in: K.R. Rao (Ed.), Pentachlorophenol: Chemistry, Pharmacology and Environmental Toxicology, Plenum, New York, pp. 325–338.

Farquharson, M.E., J.C. Gage and J. Northover (1958) The biological actions of chlorophenols, Br. J. Pharmacol., 13, 20–24.

Huff, J. (1982) Carcinogenesis bioassay results from the National Toxicology Program, Environ. Health Persp., 45, 185–198.

IARC (1979) Some halogenated hydrocarbons, IARC Monographs on the Evaluation of the Carcinogenic Risk of Chemicals to Humans, Vol. 20, pp. 195–239.

IARC (1982) Chemicals, industrial processes and industries associated with cancer in humans, IARC Monographs on the Evaluation of the Carcinogenic Risk of Chemicals to Humans, Suppl. 4, pp. 88–89.

Kinae, N., T. Hashizume, T. Makita, I. Tomita, I. Kimura and H. Kanamori (1981) Studies on the toxicity of pulp and paper mill effluents, I. Mutagenicity of the sediment samples derived from kraft paper mills, Water Res., 15, 17–24.

Lindström, K., and J. Nordin (1976) Gas chromatography–mass spectrometry of chlorophenols in spent bleach liquors, J. Chromatogr., 128, 13–26.

McMillan, S., and M. Fox (1979) Failure of caffeine to influence induced mutation frequencies and the independence of cell killing and mutation induction in V79 Chinese hamster cells, Mutation Res., 60, 91–107.

Munir, K.M., J. Nair and S.V. Bhide (1984) Comparative formation of chlorophenol metabolites from hexachlorocyclohexane in mouse and rat in vivo and in vitro, Carcinogenesis, 5, 1519–1521.

National Cancer Institute (1979) Bioassay of 2,4,6-trichlorophenol for possible carcinogenicity (Technical Report Series No. 155), DHEW Publications No. (NIH) 79-1711, U.S. Department of Health, Education, and Welfare, Washington, DC.

Nestmann, E.R., and E.G.-H. Lee (1983) Mutagenicity of constituents of pulp and paper mill effluent in growing cells of *Saccharomyces cerevisiae*, Mutation Res., 119, 273–280.

Paasivirta, J. (1978) Chlorophenols — poisons, possible environmental poisons, Kemia-Kemi, 5, 367–370.

Vogel, E., and J.L.R. Chandler (1974) Mutagenicity testing of cyclamate and some pesticides in *Drosophila melanogaster*, Experientia, 30, 621–623.

Weinbach, E.C., J. Garbus and E.C. Clagett (1965) The interaction of uncoupling phenols with mitochondria and with mitochondrial protein, J. Biol. Chem., 240, 1811–1819.

Wyllie, J.A., J. Gabica, W.W. Benson and J. Yoder (1975) Exposure and contamination of the air and employees of a pentachlorophenol plant, Pesticides Monitoring J., 9, 150–153.

Zimmermann, F.K., R.C. von Borstel, E.S. von Halle, J.M. Parry, D. Siebert, G. Zetterberg, R. Barale and N. Loprieno (1984) Testing of chemicals for genetic activity with *Saccharomyces cerevisiae*: a report of the U.S. Environmental Protection Agency Gene-Tox Program, Mutation Res., 133, 199–244.

II

Revised HK 871/93

THE TOXICITY OF CHLOROPHENOLS IN V79 CHINESE HAMSTER CELLS

Kristian Jansson and Vuokko Jansson

Department of Cell Biology, University of Jyväskylä, SF-40100 Jyväskylä, Finland

Key words: Chlorophenols; Toxicity; Cell survival; Mammalian cells, cultured; Structure-activity relationship

Correspondence to: Kristian Jansson, Department of Toxicology, National Public Health Institute, P.O. Box 95, SF-70701 Kuopio, Finland

SUMMARY

The toxicity of 9 chlorophenols to V79 Chinese hamster cells was determined in a survival assay which gives a direct measure of reproductive cell death by measuring single–cell cloning efficiency. Quantitative analyses of the structure–activity relationships showed that the toxicity was linearly related not only to hydrophobicity, as suggested by previous mammalian–cell studies, but also to electronic (electron–withdrawing) substituent effects. The extent electronic factors were related to the toxicity was assessed apart from their influence on partition and transport. The results suggest a relationship between the toxicity and the well known ability of chlorophenols to induce proton permeability in membranes.

INTRODUCTION

Chlorophenols are used widely as pesticides and as intermediates in industrial syntheses. In addition, they are formed indirectly as by-products in the chlorine bleaching of wood pulp, the chlorine disinfection of water, and the incineration of municipal waste [1]. The mode of action in the acute mammalian toxicity of chlorophenols has not been established. It has been suggested that their toxicity is associated with uncoupling of the ATP synthesis (oxidative phosphorylation) from the electron transport [2]. The chlorophenols are biologically active as uncouplers because of their ability to induce proton permeability in membranes. This protonophoric activity of chlorophenols has been related to both their hydrophobicity (lipophilicity) and proton-donating ability [3,4].

Previous studies on cultured mammalian cells have suggested that chlorophenols disrupt cell function by non-specific partitioning into hydrophobic cellular compartments rather than by being protonophores [5–7]. However, these studies have measured chlorophenol-induced alterations in metabolic pathways or structural integrity which may not necessarily be related directly to cell death. Therefore, in the present study, we determined the toxicity of chlorophenols to V79 Chinese hamster cells in a survival assay which gives a direct measure of reproductive cell death by measuring single-cell cloning efficiency, and analyzed our results quantitatively in terms of physico-chemical molecular parameters.

MATERIALS AND METHODS

Chemicals

Purified 2-, 3- and 4-chlorophenol, 2,4- and 2,6-dichlorophenol, 2,4,5- and 2,4,6-trichlorophenol, 2,3,4,6-tetrachlorophenol, and pentachlorophenol were kindly supplied by Dr. Juha Knuutinen (Department of Chemistry, University of Jyväskylä, Jyväskylä, Finland). The purity of all chlorophenols was at least 99% by gas chromatography. Stock solutions of the chlorophenols were prepared in acetone (Merck, Darmstadt, Germany) immediately before use.

Cells

V79 Chinese hamster cells, originally obtained from Dr. Eliezer Huberman (then at the Oak Ridge National Laboratory Oak Ridge, TN), were cloned, checked for the presence of mycoplasma by the method of Chen [8], and stored in ampoules frozen in liquid nitrogen. This particular subclone has a population doubling time of 14 h and a modal chromosome number of 22. Before each experiment, an ampoule of cells was thawed and maintained in exponential growth for 3 days. The cells were grown in Dulbecco's modified Eagle's medium supplemented with 10% fetal calf serum, penicillin (100 IU/ml), and streptomycin (100 μg/ml) (all from Gibco, Paisley, UK) at 37°C (pH 7.4) with a humidified atmosphere of 10% CO_2 in air. Cell-culture dishes were purchased from Nunc (Roskilde, Denmark).

Survival assay

In an experiment, 1×10^6 cells were plated in 10 ml of medium per 100-mm dish and allowed to grow for 24 h. The medium was then replaced with 10 ml of fresh medium containing the chlorophenol. Two dishes were used for each concentration of the chlorophenol and for solvent control cells. The maximum solvent concentration was 0.5%. This level had no effect on cell growth or cloning efficiency. After 24 h of treatment, the cells were washed twice with Dulbecco's phosphate buffered saline (PBS), dissociated with 0.05% trypsin and 0.02% EDTA (Gibco) in PBS, and counted in a hemacytometer. To determine single-cell cloning efficiency, 200 cells were plated in 4 ml of medium per 60-mm dish (4 dishes/concentration), and after 6 days, the colonies were fixed in ethanol and stained with Giemsa. The survival was calculated as compared to the solvent control cloning efficiency. The negative logarithm of the chlorophenol concentration required to reduce the survival to 50% of control (p*LC50*) was estimated by linear regression analysis. It was used as the parameter of toxic potency for structure-activity relationships analyzed by the Hansch method [9].

Physico-chemical parameters

The logarithm of the octanol-water partition coefficient of neutral molecules (log *P*) and the negative logarithm of the acid dissociation constant (pK_a) were taken from Ref. 10. To allow for ionization at pH 7.4, log *P* was corrected by Eqn. 1 [11]:

$$\log D = \log P - \log [1 + \text{antilog} (pH - pK_a)] \quad (1)$$

where *D* is the distribution coefficient, which takes both ionized and neutral molecules into account. The Hammett substituent constant (σ denotes the composite of σ_o (= σ_p), σ_m and σ_p) was obtained from Ref. 12.

RESULTS AND DISCUSSION

Fig. 1 shows the survival data for each of the 9 chlorophenols in the V79 cell assay. On the basis of their p$\underline{LC50}$ values listed in Table 1, the sequence of toxic potency was 2,4,5-trichlorophenol ≈ pentachlorophenol > 2,3,4,6-tetrachlorophenol > 2,4-dichlorophenol > 2,4,6-trichlorophenol > 3-chlorophenol > 2,6-dichlorophenol ≈ 4-chlorophenol > 2-chlorophenol. Regression analysis of data in Table 1 yielded Eqns. 2-4:

$$pLC50 = 0.818\ (\pm 0.492)\ \log D + 1.240\ (\pm 1.752) \quad (2)$$

$$(n = 9,\ r^2 = 0.551,\ s = 0.374)$$

$$pLC50 = 0.766\ (\pm 0.325)\ \log D - 0.215\ (\pm 0.105)\ pK_a + 2.928\ (\pm 0.871) \quad (3)$$

$$(n = 9,\ r^2 = 0.913,\ s = 0.178)$$

$$pLC50 = 0.592\ (\pm 0.141)\ \log D + 0.890\ (\pm 0.165)\ \sigma + 1.262\ (\pm 0.358) \quad (4)$$

$$(n = 9,\ r^2 = 0.985,\ s = 0.074)$$

In these expressions, n is the number of chlorophenols, r^2 is the correlation coefficient squared, and s is the standard deviation. The figures in parentheses are for construction of the 95% confidence limits. The values of p$\underline{LC50}$ predicted from Eqns. 2-4 are plotted versus the observed values in Fig. 2.

Eqns. 2-4 show that the toxicity of chlorophenols is linearly related to both hydrophobicity, log D, and electronic effects accounted for by pK_a or σ. A better correlation was obtained with σ, which represents polar effects of o-, m-and p-substituents, excluding the extra proximity effects (polar and steric) of o-substituents [13].

The use of log D instead of log P in Eqns. 3 and 4 allows the dependence of the toxicity on electronic factors to be distinguished from the effects of equilibrium ionization on partition and transport. In both equations, the log D term is proportional to the concentration of chlorophenol in the biolipid phase, while the remaining electronic effect suggests that loss of a proton is involved in the rate-limiting step of its action. The chlorophenols are known to be biologically active as uncouplers of mitochondrial oxidative phosphorylation because of their protonophoric activity, which depends on both hydrophobicity and electron-withdrawing power [3,4]. Therefore, the toxicity of chlorophenols could result from their ability to induce proton permeability in mitochondrial or other cellular membranes.

When the concentration producing a standard response (e.g., LC50, EC50) refers to that of a compound partially ionized at the pH at which the measurements have been made, log P has to be replaced by log D which takes both neutral and ionized species into account. However, it is equally correct to relate the neutral fraction of that concentration to log P. By using this method, Aschmann et al. [6] found an excellent correlation (r^2 = 0.998) between ATP depletion and hydrophobicity represented by log P for a set of chlorophenols in cultured rat hepatocytes. It is therefore tempting to conclude that electronic effects are not important in their results. However, in their data set, log P correlates strongly with both pK_a (r^2 = 0.981) and σ (r^2 = 0.988). Thus, from their data set, one cannot infer a role for both log P and electronic effects (pK_a, σ). Because of the same set of chlorophenols studied, the above collinearity problem is also inherent in the study of Seibert et al. [7], who reported the dependence of the bovine sperm cell immobilizing potency of chlorophenols on hydrophobicity represented by log P.

In conclusion, the results presented here indicate that the toxicity of chlorophenols in V79 cells depends not only on hydrophobicity but also on electron–withdrawing substituent effects. Therefore, the toxicity could be related to the ability of chlorophenols to induce proton permeability in membranes. The precise mechanism, however, remains to be investigated.

REFERENCES

1 Paasivirta, J., Heinola, K., Humppi, T., Karjalainen, A., Knuutinen, J., Mäntykoski, K., Paukku, R., Piilola, T., Surma-Aho, K., Tarhanen, J., Welling, L., Vihonen, H. and Särkkä, J. (1985) Polychlorinated phenols, guaiacols and catechols in environment. Chemosphere 14, 469–491.

2 Exon, J.H. (1984) A review of chlorinated phenols. Vet. Hum. Toxicol. 26, 508–520.

3 Stockdale, M. and Selwyn, M.J. (1971) Effects of ring substituents on the activity of phenols as inhibitors and uncouplers of mitochondrial respiration. Eur. J. Biochem. 21, 565–574.

4 Miyoshi, H., Nishioka, T. and Fujita, T. (1987) Quantitative relationship between protonophoric and uncoupling activities of substituted phenols. Biochim. Biophys. Acta 891, 194–204.

5 Babich, H. and Borenfreund, E. (1987) Structure-activity relationship (SAR) models established in vitro with the neutral red cytotoxicity assay. Toxicol. In Vitro 1, 3–9.

6 Aschmann, C., Stork, T. and Wassermann, O. (1989) Short-term effects of chlorophenols on the function and viability of primary cultured rat hepatocytes. Arch. Toxicol. 63, 121–126.

7 Seibert, H., Kolossa, M. and Wassermann, O. (1989) Bovine spermatozoa as an in vitro model for studies on the cytotoxicity of chemicals: effects of chlorophenols. Cell Biol. Toxicol. 5, 315–330.

8 Chen, T.R. (1977) In situ detection of mycoplasma contamination by fluorescent

Hoechst 33258 stain. Exp. Cell Res. 104, 255–262.

9 Hansch, C. and Fujita, T. (1969) ρ–σ–π–Analysis. A method for the correlation of biological activity and chemical structure. J. Am. Chem. Soc. 86, 1616–1626.

10 Saito, H., Sudo, M., Shigeoka, T. and Yamauchi, F. (1991) In vitro cytotoxicity of chlorophenols to goldfish GF–Scale (GFS) cells and quantitative structure–activity relationships. Environ. Toxicol. Chem. 10, 235–241.

11 Scherrer, R.A. and Howard, S.M. (1977) Use of distribution coefficients in quantitative structure–activity relationships. J. Med. Chem. 20, 53–58.

12 Hansch, C., Leo, A., Unger, S., Kim, K., Nikaitani, D. and Lien, E. (1973) "Aromatic" substituent constants for structure–activity correlations. J. Med. Chem. 16, 1207–1216.

13 Fujita, T. (1981) The ortho effect in quantitative structure–activity correlations. Anal. Chim. Acta 133, 667–676.

TABLE I

TOXIC POTENCY (p*LC50*) OF CHLOROPHENOLS IN THE V79 CELL SURVIVAL ASSAY AND PHYSICO-CHEMICAL PARAMETERS USED TO DERIVE Eqns. 2-4

Chlorophenol	p*LC50*[a]	Log *D*[b]		pK_a[c]	σ[d]
2-Chlorophenol	2.69	2.12	(2.17)	8.29	0.23
3-Chlorophenol	3.04	2.46	(2.48)	8.78	0.37
4-Chlorophenol	2.91	2.40	(2.41)	9.14	0.23
2,4-Dichlorophenol	3.63	3.13	(3.21)	8.09	0.46
2,6-Dichlorophenol	2.92	2.13	(2.84)	6.79	0.46
2,4,5-Trichlorophenol	4.05	3.60	(4.10)	7.07	0.83
2,4,6-Trichlorophenol	3.30	2.53	(3.75)	6.21	0.69
2,3,4,6-Tetrachlorophenol	3.85	2.63	(4.42)	5.62	1.06
Pentachlorophenol	4.03	2.54	(5.04)	4.90	1.43

[a]*LC50* in mol/l.

[b]By Eqn. 1. The figures in parentheses are the log *P* from Ref. 10.

[c]From Ref. 10.

[d]From Ref. 12.

FIGURE LEGENDS

Fig. 1. Survival of V79 cells after treatment for 24 h with 2,4,5-trichlorophenol (○), pentachlorophenol (●), 2,3,4,6-tetrachlorophenol (▵), 2,4-dichlorophenol (□), 2,4,6-trichlorophenol (■), 3-chlorophenol (◇), 2,6-dichlorophenol (▿), 4-chlorophenol (▴), and 2-chlorophenol (▾). Each point represents the mean of at least two independent experiments. The cloning efficiency of the solvent control cells ranged from 62 to 85%.

Fig. 2. Values of pLC50 predicted from Eqns. 2 (○), 3 (▵) and, 4 (□) versus the observed values.

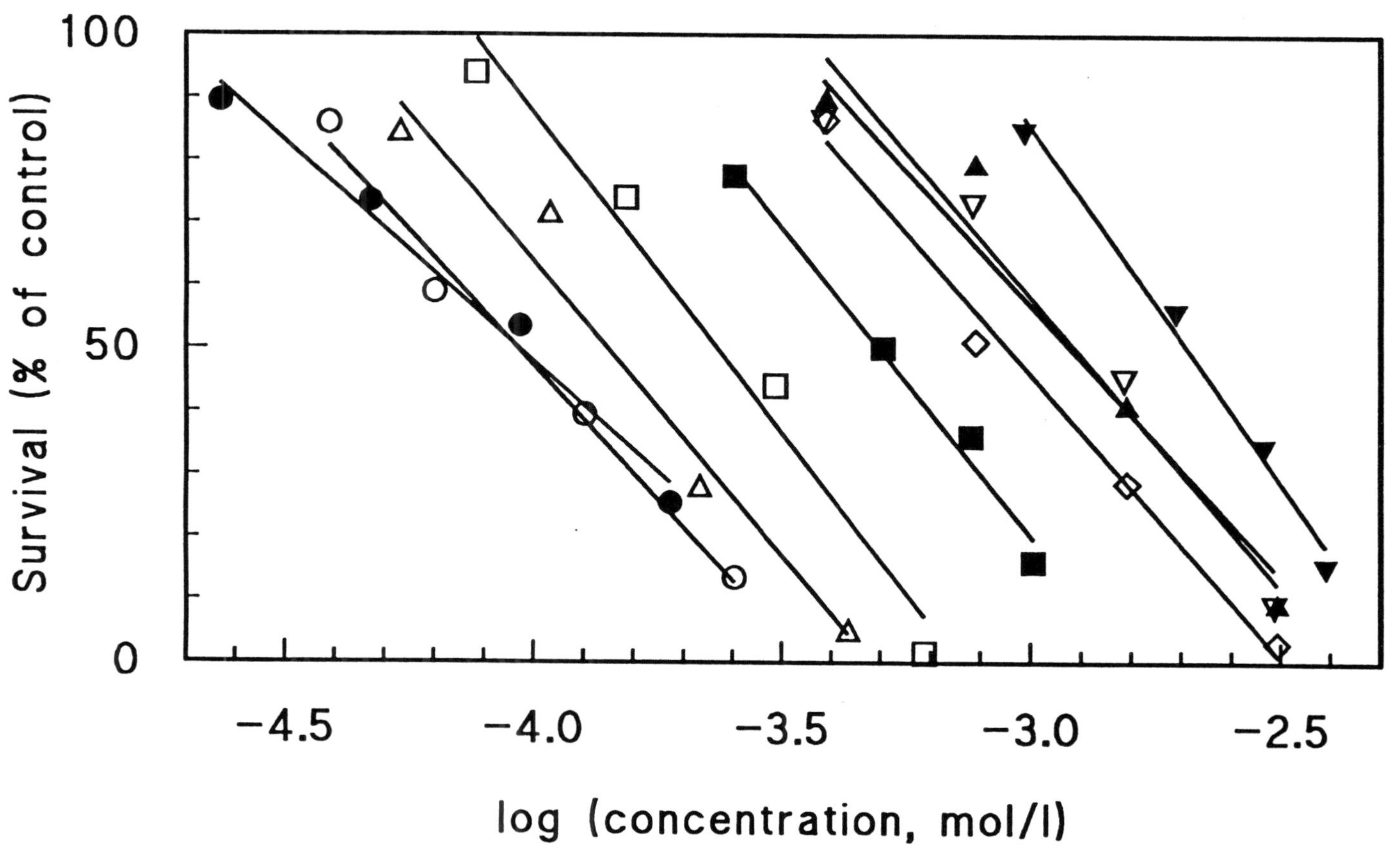

100
50
0
Survival (% of control)
−4.5
−4.0
−3.5
−3.0
−2.5
log (concentration, mol/l)

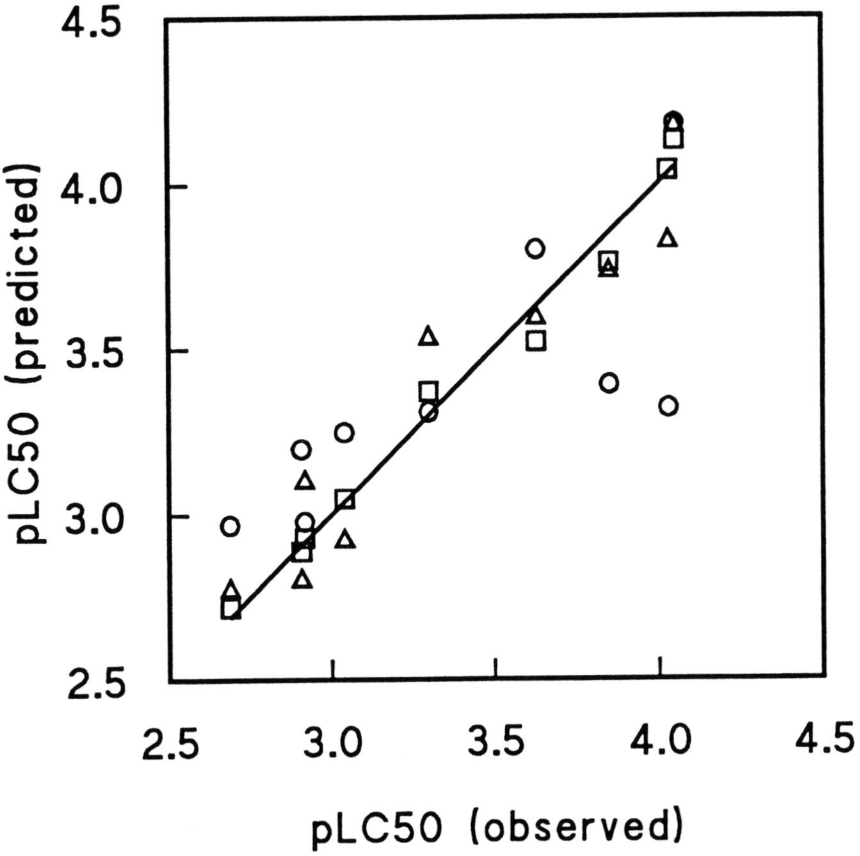

pLC50 (predicted)
4.5
4.0
3.5
3.0
2.5
2.5
3.0
3.5
4.0
4.5
pLC50 (observed)

III

Mutation Research, 280 (1992) 175–179

MUTGEN 01805

Genotoxicity of 2,4,6-trichlorophenol in V79 Chinese hamster cells

Kristian Jansson and Vuokko Jansson

Department of Cell Biology, University of Jyväskylä, SF-40100 Jyväskylä, Finland

(Received 31 December 1991)
(Revision received 25 March 1992)
(Accepted 27 March 1992)

Keywords: Chlorophenols; 2,4,6-Trichlorophenol; Carcinogens, environmental; Mutagenicity tests; Cells, cultured

Summary

The genotoxicity of the rodent carcinogen 2,4,6-trichlorophenol (TCP) was studied without exogenous metabolic activation in V79 Chinese hamster cells. TCP did not induce mutation at the *hprt* locus to 6-thioguanine resistance or structural chromosome aberrations. However, it produced statistically significant, dose-related increases in hyperdiploidy and micronuclei. From these results it appears that TCP causes chromosome malsegregation as its major mode of genotoxic action.

2,4,6-Trichlorophenol (TCP) is an environmental contaminant the main sources of which are the chlorine bleaching of wood pulp, the chlorine disinfection of waste water, and the incineration of municipal waste (Paasivirta et al., 1985). TCP is a major metabolite of the pesticide hexachlorocyclohexane in rats (Munir et al., 1984). Rats excrete TCP as unchanged TCP and TCP glucuronide in urine (Pekari et al., 1986).

There is sufficient evidence that TCP is carcinogenic in experimental animals (IARC, 1982). When given in the diet, TCP produced lymphomas and leukemias in male rats, and hepatocellular carcinomas or adenomas in both male and female mice (NCI, 1979). More recently, TCP has been found to increase the rate of tumor progression in a transplant model for leukemia in rats (Dieter et al., 1989).

TCP was not mutagenic in the Salmonella/mammalian microsome (S9) assay (Haworth et al., 1983), nor did it induce chromosome aberrations or sister-chromatid exchanges in Chinese hamster ovary (CHO) cells when tested in the presence or absence of S9 activation (Galloway et al., 1987). In L5178Y mouse lymphoma cells, however, TCP induced mutation at the *tk* locus to trifluorothymidine (TFT) resistance in the absence of exogenous metabolic activation (McGregor et al., 1988).

To extend the above studies, we tested TCP without exogenous metabolic activation in V79 Chinese hamster cells for the induction of multiple genetic endpoints including mutation at the *hprt* locus to 6-thioguanine (TG) resistance, chromosome aberrations, hyperdiploidy, and micronuclei.

Correspondence: Kristian Jansson, Department of Environmental Hygiene and Toxicology, National Public Health Institute, P.O. Box 95, SF-70701 Kuopio, Finland.

Materials and methods

Chemicals

TCP (CAS No. 88-06-2) was purchased from Fluka AG (Buchs, Switzerland), and was purified by distillation at reduced pressure to a purity of 99.7% as determined by gas chromatography. Ethyl methanesulfonate (EMS, CAS No. 62-50-0) was obtained from Sigma Chemical Co. (St. Louis, MO, USA). Stock solutions of the chemicals were prepared in dimethyl sulfoxide (E. Merck, Darmstadt, Germany) immediately before use.

Cells

V79 Chinese hamster cells, originally obtained from Dr. Eliezer Huberman (then at the Oak Ridge National Laboratory, Oak Ridge, TN, USA), were cloned to reduce the spontaneous background frequency of TG-resistant mutants, checked for the presence of mycoplasma by the method of Chen (1977), and stored in ampoules frozen in liquid nitrogen. No HAT (hypoxanthine, aminopterin, thymidine) treatments were performed. This particular subclone had a population doubling time of 14 h and a modal chromosome number of 22. Before each experiment, an ampoule of cells was thawed and maintained in exponential growth for 3 days. The cells were grown in Dulbecco's modified Eagle's medium supplemented with 10% fetal calf serum, penicillin (100 IU/ml), and streptomycin (100 μg/ml) (all from Gibco Ltd., Paisley, UK) at 37°C with a humidified atmosphere of 10% CO_2 in air. Cell culture dishes and flasks were purchased from Nunc (Roskilde, Denmark).

Cell survival and mutation

The cells were plated at 3×10^5 cells per 100-mm dish (3 dishes/dose) in 10 ml of medium and allowed to grow for 24 h. The medium was then replaced with 10 ml of fresh medium containing the test chemical. After 24 h of treatment, the cells were washed twice with Dulbecco's phosphate-buffered saline (PBS) and harvested with 0.05% trypsin and 0.02% EDTA (Gibco) in PBS. To determine cell survival, 200 cells were plated in 4 ml of medium per 60-mm dish (4 dishes/dose), and after 6 days, the colonies were fixed in ethanol and stained with Giemsa. For phenotypic expression, 2×10^6 cells were plated in a 80-cm^2 flask in 20 ml of medium. The cells were subcultured every 2 days, maintaining 2×10^6 cells at each subculture. After a 6-day expression period, the cells were plated at 1×10^5 cells per 100-mm dish (20 dishes/dose) in 10 ml of medium containing 30 μM TG (Sigma). The mutant colonies were fixed and stained 10 days later. To determine cloning efficiency at the time of mutant selection, 200 cells were plated in 4 ml of medium per 60-mm dish (4 dishes/dose), and after 6 days, the colonies were fixed and stained. The mutant frequency was expressed as mutants per 10^6 clonable cells.

Chromosome aberrations, hyperdiploidy, and micronuclei

The cells were plated at 3×10^5 cells per 100-mm dish (2 dishes/dose) in 10 ml of medium and allowed to grow for 24 h. The medium was then replaced with 10 ml of fresh medium containing the test chemical. After 24 h of treatment, the cells were washed twice with PBS, and harvested as above either immediately or after an additional 24-h incubation in fresh medium. Colcemid (Gibco Ltd.) was added to a final concentration of 0.1 μg/ml 2 h before harvest. The cells were suspended in a hypotonic solution of 75 mM KCl at 37°C for 10 min, fixed 3 times in ice-cold methanol : acetic acid (3 : 1), dropped onto wet glass slides, air-dried, and stained with 5% Giemsa in 10 mM phosphate buffer (pH 6.8). The slides were mounted in DPX and coded. For chromosome aberrations, 100 metaphase cells per dose (50 cells/slide) with 21–23 chromosomes were scored. The aberrations excluding gaps and endoreduplications were grouped into categories of 'simple' (breaks and terminal deletions), 'complex' (exchanges and rearrangements), 'other' (including pulverized chromosomes), and 'total' as described by Galloway et al. (1987). For hyperdiploidy, 200 metaphase cells per dose (100 cells/slide) were scored. Only diploid cells and hyperdiploid cells with 23–28 chromosomes were registered. For micronuclei, 2000 interphase cells per dose (1000 cells/slide) were scored using the criteria of Countryman and Heddle (1976).

Data evaluation

The variance of the mutant frequency was calculated as described by Leong et al. (1985), and a one-tailed normal deviate (z) test was used to examine increases over control levels at each dose. For chromosome aberrations ('total' category), hyperdiploidy, and micronuclei, the one-tailed z test for binomial proportions with the continuity correction was used.

Results and discussion

Multiple genetic endpoints including mutation at the *hprt* locus to TG resistance, chromosome aberrations, hyperdiploidy, and micronuclei were measured in V79 cells after treatment with TCP for 24 h in the absence of exogenous metabolic activation. Evidence that the cells were responsive to DNA damage was shown by EMS, a direct-acting alkylating agent, at 200 μg/ml.

TCP was unable to induce TG-resistant mutants at doses up to 180 μg/ml (Table 1). At this dose, the relative survival was reduced to 14%. A qualitative reduction in the size of surviving colonies occurred at 90 μg/ml and higher. Thus, at these doses, TCP affected cell proliferation as well as survival.

TCP did not induce chromosome aberrations in cells fixed immediately after treatment (Table 2). The highest dose scored was 60 μg/ml, where the mitotic index was reduced by about 78%. At this dose, TCP did not induce chromosome aberrations in cells fixed 24 h after treatment, either. These results agree with those of Galloway et al. (1987), who showed that TCP does not induce chromosome aberrations in CHO cells.

TCP produced significant, dose-related increases in hyperdiploidy and micronuclei in cells fixed 24 h after treatment (Table 3). For both responses, the lowest observed effective dose was 30 μg/ml. At doses up to 60 μg/ml, there was no reduction in the mitotic index. It appears that TCP inhibits cell proliferation by a reversible mechanism rather than by causing cell killing.

The positive results for hyperdiploidy indicate that TCP produces chromosome malsegregation leading to aneuploidy in V79 cells. Önfelt (1987) tested 4 chlorophenols (2-chlorophenol, 2,4- and 3,5-dichlorophenol, and 3,4,5-trichlorophenol) and found them all to induce spindle disturbances measured as c-mitosis in V79 cells. In addition, Önfelt (1987) found the c-mitotic activity of these chlorophenols to be related to their proton-donating ability rather than lipophilicity.

TABLE 1

INDUCTION OF 6-THIOGUANINE-RESISTANT MUTANTS IN V79 CHINESE HAMSTER CELLS BY TREATMENT WITH 2,4,6-TRICHLOROPHENOL FOR 24 h

Treatment	Relative survival (%) [a]		Mutants per 10^6 clonable cells [b]	
	Expt. 1	Expt. 2	Expt. 1	Expt. 2
DMSO (0.5%)	100	100	13 (19)	8 (13)
TCP (10 μg/ml)	109	NT	7 (9)	NT
TCP (30 μg/ml)	101	92	10 (13)	12 (18)
TCP (60 μg/ml)	76	83	7 (11)	13 (21)
TCP (90 μg/ml)	65	52	14 (18)	9 (14)
TCP (120 μg/ml)	50	51	11 (17)	9 (15)
TCP (150 μg/ml)	38	33	9 (13)	5 (7)
TCP (180 μg/ml)	NT	14	NT	7 (10)
EMS (200 μg/ml)	83	71	982 * (1097)	785 * (950)

[a] Relative survival was calculated as compared to the solvent control cloning efficiency (75% for Expt. 1, 69% for Expt. 2).
[b] Mutant frequency was calculated by dividing the number of mutant colonies (numbers in parentheses) by the number of cells plated for selection (2×10^6) corrected by the cloning efficiency at the time of selection.
NT, not tested.
* Significantly greater than the solvent control ($P < 0.05$).

TABLE 2

INDUCTION OF CHROMOSOME ABERRATIONS IN V79 CHINESE HAMSTER CELLS BY TREATMENT WITH 2,4,6-TRICHLOROPHENOL FOR 24 h

Cells were harvested immediately after treatment.

Treatment		Mitoses per 1000 cells scored	Cells with aberrations per 100 cells scored		
			Simple	Complex	Total
Expt. 1	DMSO (0.5%)	43	1	0	1
	TCP (10 μg/ml)	37	1	0	1
	TCP (30 μg/ml)	14	1	0	1
	TCP (60 μg/ml)	8	2	0	2
	TCP (90 μg/ml)	1	ND	ND	ND
	EMS (200 μg/ml)	37	6	3	9 *
Expt. 2	DMSO (0.5%)	47	2	0	2
	TCP (10 μg/ml)	41	1	1	2
	TCP (30 μg/ml)	19	3	0	3
	TCP (60 μg/ml)	12	2	1	3
	TCP (60 μg/ml)	52 [a]	1 [a]	0 [a]	1 [a]
	EMS (200 μg/ml)	31	7	5	12 *

[a] Cells were harvested 24 h after treatment.
ND, not determined.
* Significantly greater than the solvent control ($P < 0.05$).

Like all chlorophenols, TCP is a weakly acidic uncoupler of oxidative phosphorylation (Stockdale and Selwyn, 1971). It has been suggested that the protonophoric membrane action of weakly acidic uncouplers might lead to spindle disturbances in mammalian cells (see Önfelt,

TABLE 3

INDUCTION OF HYPERDIPLOIDY AND MICRONUCLEI IN V79 CHINESE HAMSTER CELLS BY TREATMENT WITH 2,4,6-TRICHLOROPHENOL FOR 24 h

Cells were harvested 24 h after treatment.

Treatment		Mitoses per 1000 cells scored	Hyperdiploid cells per 200 cells scored	Cells with micronuclei per 2000 cells scored
Expt. 1	DMSO (0.5%)	46	11	32
	TCP (10 μg/ml)	51	18	44
	TCP (30 μg/ml)	59	31 *	56 *
	TCP (60 μg/ml)	68	34 *	88 *
	TCP (90 μg/ml)	24	42 *	116 *
	EMS (200 μg/ml)	34	17	126 *
Expt. 2	DMSO (0.5%)	53	8	36
	TCP (10 μg/ml)	55	14	46
	TCP (30 μg/ml)	50	32 *	69 *
	TCP (60 μg/ml)	62	32 *	77 *
	TCP (90 μg/ml)	30	36 *	114 *
	EMS (200 μg/ml)	39	12	112 *

* Significantly greater than the solvent control ($P < 0.05$).

1986). However, the precise mechanism remains obscure.

Micronuclei arise from chromosome fragments or intact chromosomes excluded from the nucleus during cell division (Heddle et al., 1983). TCP was unable to induce chromosome aberrations. Therefore, micronucleus formation by TCP suggests malsegregational loss of whole chromosomes. TCP is a direct mutagen at the *tk* locus in L5178Y cells (McGregor et al., 1988). However, it was not mutagenic at the *hprt* locus in V79 cells. Chromosome loss could be recovered at the heterozygous autosomal *tk* locus but not at the hemizygous X-linked *hprt* locus, which permits the recovery primarily of gene mutation (cf. Liber et al., 1989).

Our present results, taken together with the fact that TCP has no structural features alerting to DNA reactivity (Ashby and Tennant, 1991), suggest that TCP causes chromosome malsegregation as its major mode of genotoxic action. It remains to be elucidated if this is a mechanism by which TCP causes cancer in experimental animals.

References

Ashby, J., and R.W. Tennant (1991) Definitive relationships among chemical structure, carcinogenicity and mutagenicity for 301 chemicals tested by the U.S. NTP, Mutation Res., 257, 229–306.

Chen, T.R. (1977) In situ detection of mycoplasma contamination by fluorescent Hoechst 33258 stain, Exp. Cell Res., 104, 255–262.

Countryman, P.I., and J.A. Heddle (1976) The production of micronuclei from chromosome aberrations in irradiated cultures of human lymphocytes, Mutation Res., 41, 321–332.

Dieter, M.P., C.W. Jameson, J.E. French, S. Gangjee, S.A. Stefanski, R.S. Chhabra and P.C. Chan (1989) Development and validation of a cellular transplant model for leukemia in Fisher rats: a short term assay for potential anti-leukemic chemicals, Leukemia Res., 13, 841–849.

Galloway, S.M., M.J. Armstrong, C. Reuben, S. Colman, B. Brown, C. Cannon, A.D. Bloom, F. Nakamura, M. Ahmed, S. Duk, J. Rimpo, B.H. Margolin, M.A. Resnick, B. Anderson and E. Zeiger (1987) Chromosome aberrations and sister chromatid exchanges in Chinese hamster ovary cells: evaluations of 108 chemicals, Environ. Mol. Mutagen., 10 (Suppl. 10), 1–175.

Haworth, S., T. Lawlor, K. Mortelmans, W. Speck and E. Zeiger (1983) Salmonella mutagenicity test results for 250 chemicals, Environ. Mutagen., 5 (Suppl. 1), 3–142.

Heddle, J.A., M. Hite, B. Kirkhart, K. Mavournin, J.T. MacGregor, G.W. Newell and M.F. Salamone (1983) The induction of micronuclei as a measure of genotoxicity, a report of the U.S. Environmental Protection Agency Gene-Tox Program, Mutation Res., 123, 61–118.

IARC (1982) Chemicals, industrial processes and industries associated with cancer in humans, IARC Monographs on the Evaluation of the Carcinogenic Risk of Chemicals to Humans, Suppl. 4, IARC, Lyon, pp. 249–250.

Leong, P.-M., W.G. Thilly and S. Morgenthaler (1985) Variance estimation in single-cell mutation assays: comparison to experimental observations in human lymphoblasts at 4 gene loci, Mutation Res., 150, 403–410.

Liber, H.L., D.W. Yandell and J.B. Little (1989) A comparison of mutation induction at the *tk* and *hprt* loci in human lymphoblastoid cells; quantitative differences are due to an additional class of mutations at the autosomal *tk* locus, Mutation Res., 216, 9–17.

McGregor, H.L., A. Brown, P. Cattanach, I. Edwards, D. McBride, C. Riach and W.J. Caspary (1988) Responses of the L5178Y tk^+/tk^- mouse lymphoma cell forward mutation assay: III. 72 coded chemicals, Environ. Mol. Mutagen., 12, 85–154.

Munir, K.M., J. Nair and S.V. Bhide (1984) Comparative formation of chlorophenol metabolites from hexachlorocyclohexane in mouse and rat in vivo and in vitro, Carcinogenesis, 5, 1519–1521.

NCI (1979) Bioassay of 2,4,6-trichlorophenol for possible carcinogenicity, Technical Report Series No. 155, DHEW Publication No. (NIH) 79-1711, DHEW, Washington, DC.

Önfelt, A. (1986) Mechanistic aspects on chemical induction of spindle disturbances and abnormal chromosome numbers, Mutation Res., 168, 249–300.

Önfelt, A. (1987) Spindle disturbances in mammalian cells, III. Toxicity, c-mitosis and aneuploidy with 22 different compounds. Specific and unspecific mechanisms, Mutation Res., 182, 135–154.

Paasivirta, J., K. Heinola, T. Humppi, A. Karjalainen, J. Knuutinen, K. Mäntykoski, R. Paukku, T. Piilola, K. Surma-Aho, J. Tarhanen, L. Welling, H. Vihonen and J. Särkkä (1985) Polychlorinated phenols, guaiacols and catechols in environment, Chemosphere, 14, 469–491.

Pekari, K., C. Boudène and A. Aitio (1986) Kinetics of 2,4,6-trichlorophenol in different organs of the rat, Arch. Toxicol., 59, 41–44.

Stockdale, M., and M.J. Selwyn (1971) Effects of ring substituents on the activity of phenols as inhibitors and uncouplers of mitochondrial respiration, Eur. J. Biochem., 21, 565–574.

IV

Mutation Research, 260 (1991) 83–87
© 1991 Elsevier Science Publishers B.V. 0165-1218/91/$03.50
ADONIS 016512189100082J

MUTGEN 01644

Induction of mutation in V79 Chinese hamster cells by tetrachlorohydroquinone, a metabolite of pentachlorophenol

Kristian Jansson and Vuokko Jansson

Department of Cell Biology, University of Jyväskylä, SF-40100 Jyväskylä (Finland)

(Received 21 June 1990)
(Revision received 5 October 1990)
(Accepted 10 October 1990)

Keywords: Chlorophenols; Pentachlorophenol; Carcinogens, environmental; Mutagenicity tests; Cells, cultured

Summary

Tetrachlorohydroquinone (TCHQ) and tetrachlorocatechol (TCC), two metabolites of the environmental mutagen and carcinogen pentachlorophenol, were tested without exogenous activation in V79 Chinese hamster cells for the induction of mutation at the hypoxanthine phosphoribosyl transferase (HPRT) locus to 6-thioguanine resistance (TG^r) and at the Na/K-ATPase locus to ouabain resistance (Oua^R). Treatment was for 24 h at 37°C.

TCHQ produced statistically significant increases in the frequency of TG^r mutants. The lowest observed effective dose (LOED) was 20 μM, where the relative cloning efficiency was 63%. The relationship between the dose of TCHQ and the frequency of TG^r mutants was approximately linear over the range of 0–60 μM with an estimated slope (±95% confidence limits) of 1.1 ± 0.3 mutants per 10^6 clonable cells per μM. At the highest tested dose of TCHQ, 60 μM, the relative cloning efficiency was reduced to 7%. In contrast to TCHQ, TCC was unable to induce TG^r mutants at doses up to 120 μM. The relative cloning efficiency at this dose was 5%. Both TCHQ and TCC were unable to induce Oua^R mutants.

The results suggest that TCHQ is at least partly responsible for the genotoxic activity of pentachlorophenol. TCHQ can produce reactive oxygen species, which may cause large genetic damage such as deletions, resulting in mutation to TG^r but not to Oua^R.

Tetrachlorohydroquinone (TCHQ), a major metabolite of the environmental contaminant pentachlorophenol (PCP) in mice and rats (Jacobson and Yllner, 1971; Ahlborg et al., 1974, 1978), binds covalently to calf thymus DNA and causes single-strand breaks in bacteriophage PM2 DNA in contrast to PCP (Witte et al., 1985). In addition to TCHQ, tetrachlorocatechol (TCC) has been identified as a metabolite of PCP in rats (Edgerton et al., 1979). Metabolism of PCP by rat liver microsomes results in formation of TCHQ and TCC, and in covalent binding to microsomal protein and to added DNA (van Ommen et al., 1986).

Correspondence: Kristian Jansson, Department of Cell Biology, University of Jyväskylä, Vapaudenkatu 4, SF-40100 Jyväskylä (Finland).

Evidently, both TCHQ and TCC are metabolites of PCP in humans (Ahlborg et al., 1974; Edgerton et al., 1979, 1981; Juhl et al., 1985).

Previous studies have shown that PCP is a mouse somatic mutagen (Fahrig et al., 1978; reviewed by Styles and Penman, 1985) but not a direct gene mutagen in V79 Chinese hamster cells (Jansson and Jansson, 1986). More recently, PCP has been shown to cause cancer in mice (National Toxicology Program, 1989), to induce chromosome aberrations in Chinese hamster ovary (CHO) cells in the presence but not the absence of S9 metabolic activation (Galloway et al., 1987), and to induce prophage lambda in *Escherichia coli* only in the presence of S9 (DeMarini et al., 1990).

In the present study, we tested TCHQ and TCC without exogenous activation in V79 cells for the induction of mutation at the hypoxanthine phosphoribosyl transferase (HPRT) locus to 6-thioguanine resistance (TG^r) and at the Na/K-ATPase locus to ouabain resistance (Oua^R).

Materials and methods

Chemicals

TCHQ (CAS No. 87-87-6, purity > 99% by titration) was purchased from Eastman Kodak Co., Rochester, NY. Ethyl methanesulfonate (EMS, CAS No. 62-50-0) was obtained from Sigma Chemical Co., St. Louis, MO. TCC (CAS No. 1198-55-6, purity > 99.9% by gas chromatography) was kindly supplied by Dr. Juha Knuutinen (for synthesis see Knuutinen, 1981). Stock solutions of the chemicals were prepared in dimethyl sulfoxide (E. Merck, Darmstadt, F.R.G.) immediately before use.

Mutation assay

V79 Chinese hamster cells, originally obtained from Dr. Eliezer Huberman (Oak Ridge National Laboratory, Oak Ridge, TN), were cloned to reduce the spontaneous background frequency of TG^r and Oua^R mutants, checked for the presence of mycoplasma by the method of Chen (1977), and stored in ampoules frozen in liquid nitrogen. No HAT (hypoxanthine, aminopterin, thymidine) treatments were performed. Before each experiment, an ampoule of cells was thawed and maintained in exponential growth for 3 days. The cells were grown in Dulbecco's modified Eagle's medium supplemented with 10% fetal calf serum, penicillin (100 IU/ml) and streptomycin (100 μg/ml) (Gibco Ltd., Paisley, Scotland) at 37°C with a humidified atmosphere of 10% CO_2 in air. Cell-culture dishes and flasks were purchased from Nunc (Roskilde, Denmark).

The mutation assay was essentially as described previously (Jansson and Jansson, 1986), but modified to detect resistance to Oua in addition to TG. In an experiment, 1×10^6 cells were plated in 10 ml of medium per 100-mm dish (2 dishes/dose) and allowed to grow for 24 h. The medium was then replaced with 10 ml of fresh medium containing the test chemical. Each experiment included a solvent control and a positive control of EMS at 1.6 mM. The final solvent concentration was 0.5% for all samples. After 24 h of treatment, the cells were rinsed with Dulbecco's phosphate-buffered saline (PBS) and dissociated with 0.05% trypsin and 0.02% EDTA (Gibco Ltd.) in PBS. To determine cloning efficiency, 200 cells were plated in 4 ml of medium per 60-mm dish (4 dishes/dose), and after 6 days, the colonies were fixed in ethanol and stained with Giemsa. For phenotypic expression, 2×10^6 cells were plated into a 80-cm^2 flask in 20 ml of medium. The cells were subcultured every 2 days, maintaining 2×10^6 cells at each subculture. After a 6-day expression period, the cells were plated at 1×10^5 cells per 100-mm dish (20 dishes/dose) in 10 ml of medium containing 30 μM TG (Sigma Chemical Co.) and at 2×10^5 cells per 100-mm dish (10 dishes/dose) in 10 ml of medium containing 1 mM Oua (Sigma Chemical Co.). The mutant colonies were fixed and stained 10 days later. To determine cloning efficiency at the time of mutant selection, 200 cells were plated in 4 ml of medium per 60-mm dish (4 dishes/dose), and after 6 days, the colonies were fixed and stained. The results were calculated as follows:

Cloning efficiency (CE)

$$= \frac{\text{number of colonies formed}}{\text{number of cells plated}}$$

$$\text{Relative CE} = \frac{\text{CE (treatment)}}{\text{CE (solvent control)}}$$

Mutant frequency (MF)

$$= \frac{\text{number of mutant colonies}}{\text{number of cells plated} \times \text{CE}}$$

MF is expressed as mutants per 10^6 clonable cells. When no mutant colonies were observed, the MF was calculated on the basis of an observation of a single mutant colony and thus represents the lower limit of detection. Each experiment was conducted twice.

Combined analyses of the 2 experiments were made. The MF data were evaluated by analysis of variance and Dunnett's test after a variance-stabilizing log transformation. Where a statistically significant ($P < 0.05$) response occurred, the lowest observed effective dose (LOED) was noted. The slope ($\pm$95% confidence limits) for the linear regression of MF on dose was calculated by the method of weighted least squares, with the weights estimated to be inversely proportional to the square of the expectation of MF, $E[\text{MF} \mid \text{dose}] = \alpha + \beta(\text{dose})$. This weighting scheme and the above log transformation are based on the assumption that for a fixed dose, the true standard deviation of MF is proportional to the expectation of MF. The relative CE is presented as the mean of the 2 experiments.

Results and discussion

Table 1 shows that treatment of V79 cells with both TCHQ and TCC resulted in dose-related decreases in CE. At a dose of 60 μM, TCHQ and TCC reduced the relative CE to 7 and 31%, respectively. For comparison, PCP reduced the relative CE of V79 cells to about 66% at the same dose (Jansson and Jansson, 1986). TCHQ produced statistically significant increases in the frequency of TG^r mutants. The LOED was 20 μM, where the relative CE was 63%. The relationship between the dose of TCHQ and the frequency of TG^r mutants was approximately linear over the range of 0–60 μM with an estimated slope ($\pm$95% confidence limits) of 1.1 ± 0.3 mutants per 10^6 clonable cells per μM. In contrast to TCHQ, TCC was unable to induce TG^r mutants at doses up to 120 μM. The relative CE at this dose was 5%. Both TCHQ and TCC were unable to induce

TABLE 1

MUTATION TO OUABAIN RESISTANCE (Oua^R) AND 6-THIOGUANINE RESISTANCE (TG^r) IN V79 CHINESE HAMSTER CELLS AFTER TREATMENT WITH TETRACHLOROHYDROQUINONE (TCHQ) AND TETRACHLOROCATECHOL (TCC) FOR 24 h

Treatment	Cloning efficiency (%) [a]		Mutant frequency [b]					
			Oua^R			TG^r		
	Expt. 1	Expt. 2	Expt. 1	Expt. 2	G.M.	Expt. 1	Expt. 2	G.M.
Solvent control	86	61	3 (4)	2 (3)	2	9 (13)	6 (8)	7
TCHQ 4 μM	75	57	1 (1)	2 (3)	1	11 (13)	7 (9)	9
20 μM	48	43	1 (1)	<1 (0)	<1	39 (45)	28 (36)	33 *
40 μM	17	23	2 (3)	1 (1)	1	50 (66)	44 (61)	47 *
60 μM	6	4	2 (3)	1 (1)	1	94 (112)	59 (67)	74 *
Positive control [c]	51	46	191 (228)	104 (117)	141 *	991 (1 184)	1 124 (1 265)	1 055 *
Solvent control	54	67	2 (3)	1 (1)	1	3 (5)	7 (9)	5
TCC 15 μM	48	49	<1 (0)	<1 (0)	<1	9 (12)	6 (8)	7
30 μM	42	47	1 (1)	1 (1)	1	5 (7)	7 (10)	6
60 μM	15	23	1 (1)	2 (2)	1	3 (4)	10 (12)	5
120 μM	2	4	<1 (0)	2 (2)	<1	<1 (0)	3 (4)	< 2
Positive control	43	46	157 (209)	180 (184)	168 *	972 (1 293)	1 469 (1 502)	1 195 *

[a] Cloning efficiency was determined immediately after treatment.

[b] Mutant frequency was calculated by dividing the number of mutant colonies (numbers in parentheses) by the number of cells plated for selection (2×10^6) corrected by the cloning efficiency at the time of selection (51–72%), and is expressed as mutants per 10^6 clonable cells. G.M., geometric mean.

[c] Ethyl methanesulfonate at 1.6 mM.

* Significantly different from the concurrent solvent control ($P < 0.05$).

Oua^R mutants. At 1.6 mM, EMS (positive control) produced more TG^r than Oua^R mutants per 10^6 clonable cells. The difference is consistent with previous observations (Fox et al., 1982).

In contrast to TCC, TCHQ readily autoxidizes at physiological pH values (Pritsos et al., 1987). As a result, TCHQ is in redox equilibrium with the corresponding semiquinone radical and benzoquinone, the former of which is likely to be responsible for the ability of TCHQ to bind covalently to DNA (Witte et al., 1985; van Ommen et al., 1986). On the other hand, the autoxidation of TCHQ produces reactive oxygen species, which have been shown to be responsible for its ability to induce DNA-strand breaks (Witte et al., 1985; Carstens et al., 1990). In mammalian cells, the genotoxic effects of ionizing radiation and of radiomimetic chemicals such as bleomycin and streptonigrin are evidently mediated through reactive oxygen species, which cause DNA strand breaks leading to chromosome aberrations and deletion mutations (Hsie et al., 1986; Oya et al., 1986). Like ionizing radiation (van Zeeland and Simons, 1976; Thacker et al., 1978), TCHQ was unable to induce Oua^R mutants and, at approximately equitoxic doses, was much less efficient than EMS in inducing TG^r mutants (Table 1). Note that the peripheral lymphocytes of PCP-exposed workers showed induced chromosome aberrations qualitatively similar to those after exposure to ionizing radiation (Bauchinger et al., 1982).

In the present study, we found that TCHQ but not TCC induced gene mutations in V79 cells when tested without exogenous metabolic activation. The results suggest that TCHQ is at least partly responsible for the genotoxic activity of PCP. Presumably, the mutagenic action of TCHQ is mediated through reactive oxygen species, which may cause large genetic damage such as deletions, resulting in mutation to TG^r but not to Oua^R. The clastogenicity of TCHQ remains to be evaluated.

References

Ahlborg, U.G., J.-E. Lindgren and M. Mercier (1974) Metabolism of pentachlorophenol, Arch. Toxicol., 32, 271–281.

Ahlborg, U.G., K. Larsson and T. Thunberg (1978) Metabolism of pentachlorophenol in vivo and in vitro, Arch. Toxicol., 40, 45–53.

Bauchinger, M., J. Dresp, E. Schmid and R. Hauf (1982) Chromosome changes in lymphocytes after occupational exposure to pentachlorophenol (PCP), Mutation Res., 102, 83–88.

Carstens, C.-P., J.K. Blum and I. Witte (1990) The role of hydroxyl radicals in tetrachlorohydroquinone induced DNA strand break formation in PM2 DNA and human fibroblasts, Chem.-Biol. Interact., 74, 305–314.

Chen, T.R. (1977) In situ detection of mycoplasma contamination by fluorescent Hoechst 33258 stain, Exp. Cell Res., 104, 255–262.

DeMarini, D.M., H.G. Brooks and D.G. Parkes Jr. (1990) Induction of prophage lambda by chlorophenols, Environ. Mol. Mutagen., 15, 1–9.

Edgerton, T.R., R.F. Moseman, R.E. Linder and L.H. Wright (1979) Multi-residue method for the determination of chlorinated phenol metabolites in urine, J. Chromatogr., 170, 331–342.

Edgerton, T.R., R.F. Moseman and L.H. Wright (1981) Method development for the assessment of possible human exposure to pesticides, EPA Report No. 600/1-81-024, U.S. Environmental Protection Agency, Washington, DC.

Fahrig, R., C.-A. Nilsson and C. Rappe (1978) Genetic activity of chlorophenols and chlorophenol impurities, Environ. Sci. Res., 12, 325–338.

Fox, M., S. McMillan, L. Durrant and J.M. Boyle (1982) Relative sensitivity of V79 and V79/79 cells to spontaneous and induced mutation to 6-thioguanine and ouabain resistance, Mutation Res., 95, 339–352.

Galloway, S.M., M.J. Armstrong, C. Reuben, S. Colman, B. Brown, C. Cannon, A.D. Bloom, F. Nakamura, M. Ahmed, S. Duk, J. Rimpo, B.H. Margolin, M.A. Resnick, B. Anderson and E. Zeiger (1987) Chromosome aberrations and sister chromatid exhanges in Chinese hamster ovary cells: evaluations of 108 chemicals, Environ. Mol. Mutagen., 10 (Suppl. 10), 1–175.

Hsie, A.W., L. Recio, D.S. Katz, C.Q. Lee and M. Wagner (1986) Evidence for reactive oxygen species inducing mutations in mammalian cells, Proc. Natl. Acad. Sci. (U.S.A.), 83, 9616–9620.

Jacobson, I., and S. Yllner (1971) Metabolism of ^{14}C-pentachlorophenol in the mouse, Acta Pharmacol. Toxicol., 29, 513–524.

Jansson, K., and V. Jansson (1986) Inability of chlorophenols to induce 6-thioguanine-resistant mutants in V79 Chinese hamster cells, Mutation Res., 171, 165–168.

Juhl, U., I. Witte and W. Butte (1985) Metabolism of pentachlorophenol to tetrachlorohydroquinone by human liver homogenate, Bull. Environ. Contam. Toxicol., 35, 596–601.

Knuutinen, J. (1981) Thin-layer chromatography of chlorinated catechols on a silica gel 60 layer, J. Chromatogr., 209, 446–450.

National Toxicology Program (1989) Toxicology and carcinogenesis studies of two pentachlorophenol technical-grade mixtures (CAS No. 87-86-5) in B6C3F$_1$ mice (feed studies),

NTP Technical Report No. 349, U.S. National Toxicology Program, Research Triangle Park, NC.

Oya, Y., K. Yamamoto and A. Tonomura (1986) The biological activity of hydrogen peroxide, I. Induction of chromosome-type aberrations susceptible to inhibition by scavengers of hydroxyl radicals in human embryonic fibroblasts, Mutation Res., 172, 245–253.

Pritsos, C.A., M. Pointon and R.S. Pardini (1987) Interaction of chlorinated phenolics and quinones with the mitochondrial respiration: a comparison of the *o*- and *p*-chlorinated quinones and hydroquinones, Bull. Environ. Contam. Toxicol., 38, 847–855.

Styles, J.A., and M.G. Penman (1985) The mouse spot test, Evaluation of its performance in identifying chemical mutagens and carcinogens, Mutation Res., 154, 183–204.

Thacker, J., M.A. Stephens and A. Stretch (1978) Mutation to ouabain resistance in Chinese hamster cells: induction by ethyl methanesulfonate and lack of induction by ionising radiation, Mutation Res., 51, 255–270.

van Ommen, B., A. Adang, F. Müller and P.J. van Bladeren (1986) The microsomal metabolism of pentachlorophenol and its covalent binding to protein and DNA, Chem.-Biol. Interact., 60, 1–11.

van Zeeland, A.A., and J.W.I.M. Simons (1976) Linear dose-response relationships after prolonged expression times in V-79 Chinese hamster cells, Mutation Res., 35, 129–138.

Witte, I., U. Juhl and W. Butte (1985) DNA-damaging properties and cytotoxicity in human fibroblasts of tetrachlorohydroquinone, a pentachlorophenol metabolite, Mutation Res., 145, 71–75.

V

Mutation Research, 279 (1992) 205–208

MUTGEN 01767

Induction of micronuclei in V79 Chinese hamster cells by tetrachlorohydroquinone, a metabolite of pentachlorophenol

Kristian Jansson and Vuokko Jansson

Department of Cell Biology, University of Jyväskylä, SF-40100 Jyväskylä (Finland)

(Received 6 September 1991)
(Accepted 26 November 1991)

Keywords: Pentachlorophenol; Tetrachlorohydroquinone; Hydroxyl radicals; Micronuclei; V79 cells

Summary

Tetrachlorohydroquinone, a metabolite of the fungicide pentachlorophenol, induced significant dose-related increases in micronuclei in V79 Chinese hamster cells without exogenous metabolic activation. The lowest observed effective dose was 10 μM, where the relative survival was about 62%. At the highest dose tested, 20 μM, the relative survival was about 8% and the frequency of cells with micronuclei was about 6 times the solvent control frequency. The induction of micronuclei by tetrachlorohydroquinone was significantly inhibited by the hydroxyl radical scavenger dimethyl sulfoxide at 5% (v/v).

Tetrachlorohydroquinone (TCHQ) has been identified as a major metabolite of the fungicide pentachlorophenol (PCP) in mice and rats (Jacobson and Yllner, 1971; Ahlborg et al., 1974, 1978). TCHQ has also been detected in the urine of humans occupationally exposed to PCP (Ahlborg et al., 1974; Edgerton et al., 1979). Both rat and human liver S9 fractions metabolize PCP to TCHQ (Juhl et al., 1985).

PCP has been shown by Bauchinger et al. (1982) to be a human clastogen (reviewed by Seiler, 1991). More recently, PCP has been shown to cause cancer in mice (National Toxicology Program, 1989), to induce chromosome aberrations in Chinese hamster ovary (CHO) cells in the presence but not the absence of S9 metabolic activation (Galloway et al., 1987), and to induce prophage lambda in *Escherichia coli* only in the presence of S9 (DeMarini et al., 1990).

When tested in the absence of exogenous metabolic activation, TCHQ, but not PCP, induced mutations at the *hprt* locus in V79 Chinese hamster cells (Jansson and Jansson, 1986, 1991) and DNA single-strand breaks in CHO cells (Ehrlich, 1990); TCHQ also induced DNA single-strand breaks in human fibroblasts (Carstens et al., 1990). These results suggest that TCHQ is at least partly responsible for the genotoxic activity of PCP.

To extend the above studies, we tested TCHQ without exogenous metabolic activation in V79 cells for the induction of chromosomal damage measured as micronuclei.

Correspondence: Kristian Jansson, Department of Environmental Hygiene and Toxicology, National Public Health Institute, P.O. Box 95, SF-70701 Kuopio (Finland).

Materials and methods

Chemicals

TCHQ (CAS No. 87-87-6, purity > 99% by titration) was purchased from Eastman Kodak Co. (Rochester, NY). Ethyl methanesulfonate (EMS, CAS No. 62-50-0) was obtained from Sigma Chemical Co. (St. Louis, MO). Stock solutions of the chemicals were prepared in dimethyl sulfoxide (DMSO, CAS No. 67-68-5; E. Merck, Darmstadt, Germany) immediately before use.

Cells

V79 Chinese hamster cells, originally obtained from Dr. Eliezer Huberman (then at the Oak Ridge National Laboratory, Oak Ridge, TN), were cloned, checked for the presence of mycoplasma by the method of Chen (1977), and stored in ampoules frozen in liquid nitrogen. This particular subclone had a population doubling time of 14 h and a modal chromosome number of 22. Before each experiment, an ampoule of cells was thawed and maintained in exponential growth for 3 days. The cells were grown in Dulbecco's modified Eagle's medium supplemented with 10% fetal calf serum, penicillin (100 IU/ml) and streptomycin (100 μg/ml) (all from Gibco Ltd., Paisley, U.K.) at 37°C in a humidified gas phase of 10% CO_2 in air. Cell culture dishes were purchased from Nunc (Roskilde, Denmark).

Treatment

In an experiment, 3×10^5 cells were plated in 10 ml of medium per 100-mm dish (2 dishes/dose) and allowed to grow for 24 h. The cells (approximately 1×10^6 cells per dish) were treated for 3 h at 37°C in 10 ml of serum-free Dulbecco's modified Eagle's medium (pH 7.4) without antibiotics.

Cell survival

After treatment, the cells were washed twice with Dulbecco's phosphate-buffered saline (PBS) and dissociated with 0.05% trypsin and 0.02% EDTA (Gibco Ltd.) in PBS. To determine cloning efficiency, 200 cells were plated in 4 ml of medium per 60-mm dish (4 dishes/dose), and after 6 days, the colonies were fixed in ethanol and stained with Giemsa. The relative survival was calculated as compared to the solvent control cloning efficiency.

Micronuclei

After treatment, the cells were washed twice with PBS, and 10 ml of medium was added. After an additional 24-h incubation, the cells were dissociated as above, treated with hypotonic 75 mM KCl for 10 min at 37°C, fixed 3 times in ice-cold methanol : acetic acid (3 : 1), dropped onto wet glass slides, air-dried, and stained with 5% Giemsa in 10 mM phosphate buffer (pH 6.8) for 20 min. The slides were mounted in DPX, coded, and 2000 interphase cells per dose (1000 cells/slide) were scored for micronuclei using the criteria of Countryman and Heddle (1976). The data were evaluated using the chi-square test with Yates' correction.

Results and discussion

TCHQ produced a shouldered survival curve with a D_{50} (dose to reduce relative survival to 50%) of about 12 μM in V79 cells. This curve, shown in Fig. 1, is very similar to that reported by Witte et al. (1985) in human fibroblasts. For

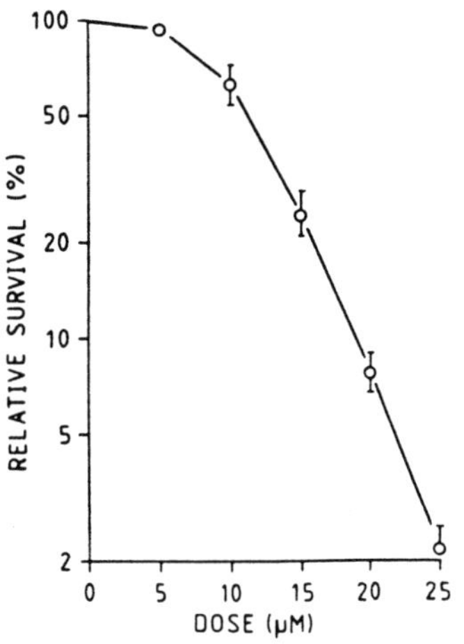

Fig. 1. Relative survival of V79 cells after treatment with TCHQ for 3 h at 37°C in serum-free medium (pH 7.4). Standard errors of the means for 3–4 separate experiments are indicated when larger than the symbol. Solvent control (0.1% DMSO) cloning efficiencies ranged from 60 to 72%.

TABLE 1

INDUCTION OF MICRONUCLEI BY TCHQ IN V79 CELLS

Treatment [a] (μM)	Cells with micronuclei per 2000 cells scored [b]	
	Expt. 1	Expt. 2
Solvent control (0.1% DMSO)	29	34
Positive control (12 mM EMS)	121 *	143 *
TCHQ		
5	53	45
10	89 *	110 *
15	142 *	151 *
20	195 *	163 *

[a] Treatment was for 3 h at 37°C in serum-free medium (pH 7.4).
[b] Cells were harvested 24 h after treatment.
* Significantly different from the solvent control ($P < 0.01$).

micronucleus experiments, we selected 4 doses that allowed relative survival of about 8 to 94%.

Table 1 shows that TCHQ produced significant dose-related increases in micronuclei. In both experiments, the lowest effective dose was 10 μM, where the relative survival was about 62%. At the highest dose tested, 20 μM, the average frequency of cells with micronuclei was about 6 times the solvent control frequency. These results agree with previous studies showing that TCHQ induces DNA single-strand breaks in CHO cells (Ehrlich, 1990) and mutations at the *hprt* locus in V79 cells (Jansson and Jansson, 1991).

TABLE 2

INHIBITORY EFFECT OF DMSO ON THE INDUCTION OF MICRONUCLEI BY TCHQ IN V79 CELLS

Treatment [a]	Cells with micronuclei per 2000 cells scored [b]	
	Expt. 1	Expt. 2
DMSO (0.1%)	32	37
DMSO (5%)	26	34
EMS (12 mM) + DMSO (0.1%)	124	131
EMS (12 mM) + DMSO (5%)	109	138
TCHQ (15 μM) + DMSO (0.1%)	133	145
TCHQ (15 μM) + DMSO (5%)	69 *	84 *

[a] Treatment was for 3 h at 37°C in serum-free medium (pH 7.4).
[b] Cells were harvested 24 h after treatment.
* Significantly different from the corresponding treatment with 0.1% DMSO ($P < 0.01$).

TCHQ readily autoxidizes at physiological pH values to the corresponding semiquinone radical (Pritsos et al., 1987; Carstens et al., 1990). As a result, reactive oxygen species are formed. Recently, Carstens et al. (1990) showed that the hydroxyl radical scavenger DMSO at 5% effectively reduces the number of TCHQ-induced DNA single-strand breaks in human fibroblasts. It was therefore of interest to us to investigate if hydroxyl radicals were involved in the formation of micronuclei in V79 cells by TCHQ.

As shown in Table 2, the induction of micronuclei by TCHQ was significantly inhibited by DMSO at 5%. This DMSO concentration has previously been shown to protect V79 cells against the formation of DNA single-strand breaks by hydroxyl radicals from hydrogen peroxide (Ward et al., 1985). As expected, DMSO had no effect on the induction of micronuclei by EMS, an alkylating agent (Table 2). This confirms that the inhibition of TCHQ-induced micronucleus formation by DMSO was due to its scavenging action.

From our present results, we conclude that TCHQ is a potent inducer of chromosomal damage in V79 cells, and that hydroxyl radicals are at least partly responsible for this damage. As a major metabolite of PCP, TCHQ should be taken into consideration in the evaluation of the genotoxic and carcinogenic potential of PCP.

References

Ahlborg, U.G., J.-E. Lindgren and M. Mercier (1974) Metabolism of pentachlorophenol, Arch. Toxicol., 32, 271–281.

Ahlborg, U.G., K. Larsson and T. Thunberg (1978) Metabolism of pentachlorophenol in vivo and in vitro, Arch. Toxicol., 40, 45–53.

Bauchinger, M., J. Dresp, E. Schmid and R. Hauf (1982) Chromosome changes in lymphocytes after occupational exposure to pentachlorophenol (PCP), Mutation Res., 102, 83–88.

Carstens, C.-P., J.K. Blum and I. Witte (1990) The role of hydroxyl radicals in tetrachlorohydroquinone induced

DNA strand break formation in PM2 DNA and human fibroblasts, Chem.-Biol. Interact., 74, 305–314.

Chen, T.R. (1977) In situ detection of mycoplasma contamination by fluorescent Hoechst 33258 stain, Exp. Cell Res., 104, 255–262.

Countryman, P.I., and J.A. Heddle (1976) The production of micronuclei from chromosome aberrations in irradiated cultures of human lymphocytes, Mutation Res., 41, 321–332.

DeMarini, D.M., H.G. Brooks and D.G. Parkes Jr. (1990) Induction of prophage lambda by chlorophenols, Environ. Mol. Mutagen., 15, 1–9.

Edgerton, T.R., R.F. Moseman, R.E. Linder and L.H. Wright (1979) Multi-residue method for the determination of chlorinated phenol metabolites in urine, J. Chromatogr., 170, 331–342.

Ehrlich, W. (1990) The effect of pentachlorophenol and its metabolite tetrachlorohydroquinone on cell growth and the induction of DNA damage in Chinese hamster ovary cells, Mutation Res., 244, 299–302.

Galloway, S.M., M.J. Armstrong, C. Reuben, S. Colman, B. Brown, C. Cannon, A.D. Bloom, F. Nakamura, M. Ahmed, S. Duk, J. Rimpo, B.H. Margolin, M.A. Resnick, B. Anderson and E. Zeiger (1987) Chromosome aberrations and sister chromatid exchanges in Chinese hamster ovary cells: evaluations of 108 chemicals, Environ. Mol. Mutagen., 10 (Suppl. 10), 1–175.

Jacobson, I., and S. Yllner (1971) Metabolism of ^{14}C-pentachlorophenol in the mouse, Acta Pharmacol. Toxicol., 29, 513–524.

Jansson, K., and V. Jansson (1986) Inability of chlorophenols to induce 6-thioguanine-resistant mutants in V79 Chinese hamster cells, Mutation Res., 171, 165–168.

Jansson, K., and V. Jansson (1991) Induction of mutation in V79 Chinese hamster cells by tetrachlorohydroquinone, a metabolite of pentachlorophenol, Mutation Res., 260, 83–87.

Juhl, U., I. Witte and W. Butte (1985) Metabolism of pentachlorophenol to tetrachlorohydroquinone by human liver homogenate, Bull. Environ. Contam. Toxicol., 35, 596–601.

National Toxicology Program (1989) Toxicology and carcinogenesis studies of two pentachlorophenol technical-grade mixtures (CAS No. 87-86-5) in $B6C3F_1$ mice (feed studies), NTP Technical Report No. 349, U.S. National Toxicology Program, Research Triangle Park, NC.

Pritsos, C.A., M. Pointon and R.S. Pardini (1987) Interaction of chlorinated phenolics and quinones with the mitochondrial respiration: a comparison of the *o*- and *p*-chlorinated quinones and hydroquinones, Bull. Environ. Contam. Toxicol., 38, 847–855.

Seiler, J.P. (1991) Pentachlorophenol, Mutation Res., 257, 27–47.

Ward, J.F., W.F. Blakely and E.I. Joner (1985) Mammalian cells are not killed by DNA single-strand breaks caused by hydroxyl radicals from hydrogen peroxide, Radiat. Res., 103, 383–392.

Witte, I., U. Juhl and W. Butte (1985) DNA-damaging properties and cytotoxicity in human fibroblasts of tetrachlorohydroquinone, a pentachlorophenol metabolite, Mutation Res. 145, 71–75.